Op-amps Circuits Manual

Newnes Circuits Manual Series

Audio IC Circuits Manual   R.M. Marston
CMOS Circuits Manual   R.M. Marston
Diode, Transistor & FET Circuits Manual   R.M. Marston
Electronic Alarm Circuits Manual   R.M. Marston
Instrumentation & Test Gear Circuits Manual   R.M. Marston
Op-amp Circuits Manual   R.M. Marston
Optoelectronics Circuits Manual   R.M. Marston
Power Control Circuits Manual   R.M. Marston
Timer/Generator Circuits Manual   R.M. Marston

Op-amp Circuits Manual

Including OTA circuits

R. M. Marston

NEWNES

To Alex, with best wishes

Butterworth-Heinemann Ltd
Linacre House, Jordan Hill, Oxford OX2 8DP

A member of the Reed Elsevier group

OXFORD   MUNICH   BOSTON
MUNICH   NEW DELHI   SINGAPORE   SYDNEY
TOKYO   TORONTO   WELLINGTON

First published 1989
Reprinted 1993

© R. M. Marston 1989

All rights reserved. No part of this publication
may be reproduced in any material form (including
photocopying or storing in any medium by electronic
means and whether or not transiently or incidentally
to some other use of this publication) without the
written permission of the copyright holder except in
accordance with the provisions of the Copyright,
Designs and Patents Act 1988 or under the terms of a
licence issued by the Copyright Licensing Agency Ltd,
90 Tottenham Court Road, London, England W1P 9HE.
Applications for the copyright holder's written
permission to reproduce any part of this publication
should be addressed to the publishers

British Library Cataloguing in Publication Data
Marston, R. M. (Raymond Michael) 1937–
  Op-amps circuits manuals
  1. Electronic equipment – Integrated circuits.
  Operational amplifiers
  I. Title
  621.381'735

ISBN 0 7506 0998 2

Printed and bound in Great Britain by Redwood Books,
Trowbridge, Wiltshire

Contents

Preface

1	Basic principles and configurations	1
2	Amplifiers and active filters	19
3	Voltage comparator circuits	38
4	Waveform generator circuits	
5	Instrumentation and test gear circuits	68
6	Compound circuits	84
7	Norton op-amp circuits	102
8	CA3080 OTA circuits	130
9	LM13600 OTA circuits	146
10	LM10 op-amp/reference circuits	168
Index		207

Preface

The operational amplifier (op-amp) is a direct-coupled high-gain differential amplifier that can readily be used as the basis of a variety of ac or dc amplifiers, instrumentation circuits, oscillators, tone generators and sensing circuits. It is one of the most popular and versatile 'building blocks' used in modern electronic circuit design and is available in three basic forms – the standard (741, etc.) type, the Norton (LM3900, etc.) type, and the operational transductance amplifier (OTA) (CA3080 and LM13600, etc.) types. This book explains how these devices work and shows how to use them in practical applications.

The book is divided into ten chapters and presents a total of over 300 practical circuits, diagrams and tables. The first six chapters deal with the operating principles and applications of the standard op-amp of the 741 (etc.) type. The remaining four chapters deal with special types of op-amp, such as the Norton amplifier (Chapter 7), the OTA (Chapters 8 and 9), and the LM10 op-amp/reference IC (Chapter 10).

The book is specifically aimed at the practical design engineer, technician and experimenter, but will be of equal interest to the electronics student and the amateur. It deals with its subject in an easy-to-read, down-to-earth, non-mathematical but very comprehensive manner. Each chapter starts off by explaining the basic principles of its subject and then goes on to present the reader with a wide range of practical circuit designs.

Throughout the book great emphasis is placed on practical 'user' information and circuitry, and the book abounds with useful circuits and data. All of the ICs and other devices used in the practical circuits are modestly priced and readily available types, with universally recognized type numbers.

R. M. Marston

1 Basic principles and configurations

An operational amplifier (op-amp) can be simply described as a high-gain direct-coupled voltage amplifier 'block' with a single output terminal but with both inverting and non-inverting input terminals, enabling the device to function as either an inverting, non-inverting, or differential amplifier. Op-amps are very versatile devices. When coupled to suitable feedback networks they can be used to make precision ac and dc amplifiers and filters, oscillators, and level switches and comparators, etc. Circuits of all these types are described in this volume.

Three basic types of operational amplifier are currently available. The most important and best known of these is the conventional voltage-in voltage-out op-amp (typified by the 741 and 3140, etc.), and the first six chapters of this volume are devoted to looking at the operating principles and applications of this type of device.

The other two basic types of op-amp are the current-differencing or Norton op-amp (typified by the LM3900 quad amplifier integrated circuit), which is particularly useful in applications using single-ended power supplies, and the voltage-in current-out variable-gain operational transconductance amplifier (OTA) (typified by the CA3080 and LM13600 ICs), which is particularly useful in voltage-controlled amplifier (VCA), voltage-controlled filter (VCF), and voltage-controlled oscillator (VCO) applications. Chapters 7 to 9 give detailed descriptions of these two families of op-amps.

Finally, one unique member of the op-amp family is an integrated circuit (IC) known as the LM10, which houses a special op-amp and a precision voltage reference in an 8-pin TO5 package; this device is fully described, complete with sixty-five 'application' circuits, in Chapter 10.

2 Basic principles and configurations

Op-amp basics

In its simplest form, a conventional op-amp consists of a differential amplifier (bipolar or FET) followed by offset compensation and output stages (as shown in *Figure 1.1*), all integrated on to a single chip and housed in an IC package. The differential amplifier has a high-impedance (constant-current) tail, to give a high-input impedance and good common-mode signal rejection to its two input terminals, and has a high-impedance collector (or drain) load, to give high signal-voltage gain (typically about 100 dB).

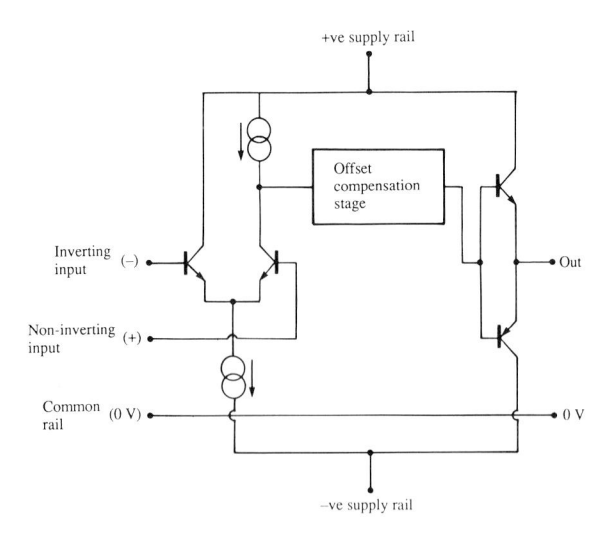

Figure 1.1 *Simplified op-amp equivalent circuit*

The output of the differential amplifier is fed to the circuit's output stage via an offset compensation network which, when the op-amp is suitably powered, causes the output to centre on zero volts when both input terminals are tied to zero volts. The output stage takes the form of a complementary emitter follower, and gives a low-impedance output.

Op-amps are represented by the standard symbol shown in *Figure 1.2(a)*. They are normally powered from split supplies, as shown in *Figure 1.2(b)*, providing positive, negative and common (zero volt) supply rails, enabling the op-amp output to swing either side of the zero volts value and to be set at zero when the differential input voltage is zero. They can, however, also be powered from single-ended supplies if required.

Basic principles and configurations 3

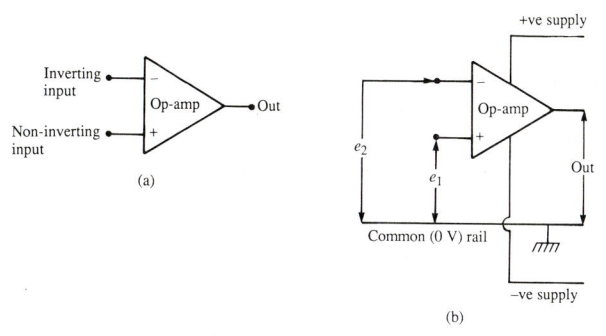

Figure 1.2 *(a) Basic symbol and (b) supply connections of an op-amp*

Basic configurations

The output signal voltage of an op-amp is proportional to the *differential* signal voltage between its two input terminals and, at low audio frequencies, is given by:

$e_{out} = A_o(e_1 - e_2)$

where A_o is the low frequency open-loop voltage gain of the op-amp (typically 100 dB, or × 100,000, e_1 is the signal voltage at the non-inverting input terminal, and e_2 is the signal voltage at the inverting input terminal.

Thus, an op-amp can be used as a high-gain inverting ac amplifier by grounding its non-inverting terminal and feeding the input signal to its inverting terminal via C_1 and R_1, as shown in *Figure 1.3(a)*, or it can be used as a non-inverting ac amplifier by reversing the two input connections as shown in *Figure 1.3(b)*, or as a differential amplifier by feeding the two input signals to the op-amp as shown in *Figure 1.3(c)*. Note in the latter case that if identical signals are fed to both input terminals the op-amp should, ideally, give zero signal output.

The voltage gains of the *Figure 1.3* circuits depend on the individual op-amp open-loop voltage gains and on the actual frequencies of the input signals. *Figure 1.4* shows the typical frequency response graph of the well known type 741 op-amp. Note that the device gives a low-frequency (below 10 Hz) voltage gain of 106 dB, but that the gain falls off at a 6 dB/octave (20 dB/decade) rate at frequencies above 10 Hz, and eventually falls to unity (0 dB) at an f_T 'unity gain transition' frequency of 1 MHz. This graph is typical of most modern op-amps, although individual op-amp types may offer different value of low-frequency gain and f_T.

One special application of the 'open-loop' op-amp is as a differential voltage comparator, one version of which is shown in *Figure 1.5(a)*. Here, a

4 Basic principles and configurations

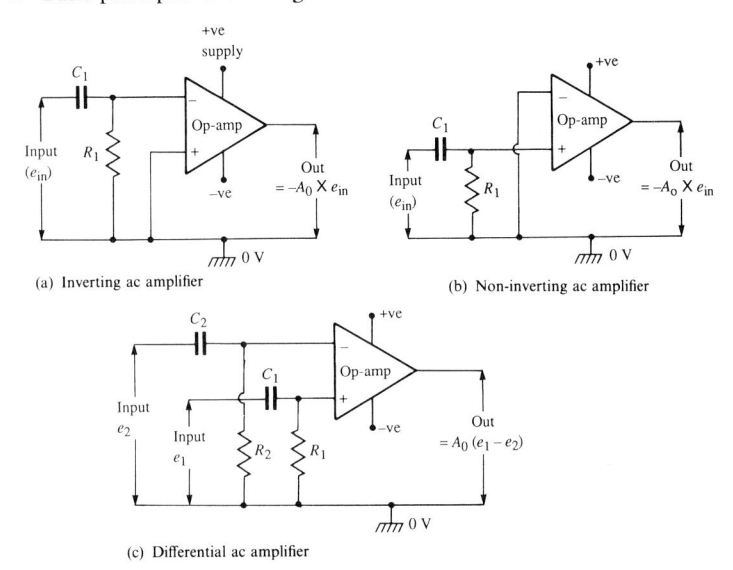

Figure 1.3 *Methods of using the op-amp as a high-gain open-loop ac amplifier*

(a) Inverting ac amplifier

(b) Non-inverting ac amplifier

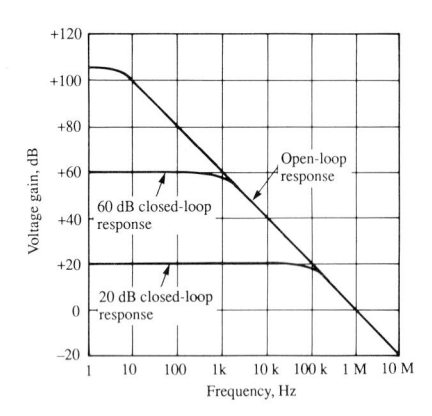

(c) Differential ac amplifier

Figure 1.4 *Typical frequency response curve of the 741 op-amp*

fixed reference voltage is applied to the inverting terminal and a variable test or sample voltage is fed to the non-inverting terminal. Because of the very high open-loop voltage gain of the op-amp, the output is driven to positive saturation (close to the positive rail value) when the sample voltage is more than a few hundred microvolts above the reference voltage, and to negative

Basic principles and configurations 5

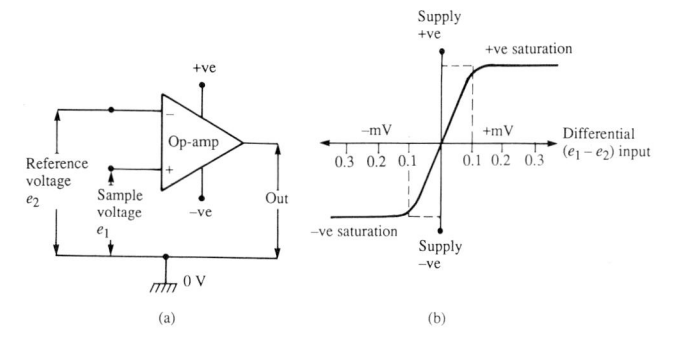

Figure 1.5 (a) Circuit and (b) transfer characteristics of a simple differential voltage comparator

saturation (close to the negative supply rail value) when the sample is more than a few hundred microvolts below the reference value.
Figure 1.5(b) shows the voltage transfer characteristics of the above circuit. Note that the output voltage magnitude is determined by the magnitude of the *differential* input voltage; thus, if a 2 V reference is used and a differential voltage of only 200 μV is needed to swing the output from a negative to a positive saturation level, this change can be caused by a shift of only 0.01% on a 2 V signal applied to the sample input. This circuit thus functions as a precision voltage comparator or balance detector.

Closed-loop amplifiers

The best way of using an op-amp as an ac amplifier is to connect it in the closed-loop mode, with negative feedback applied from output to input as shown in the circuits of *Figure 1.6*, so that the overall gain of each circuit is precisely determined by the values of the external feedback components almost irrespective of the individual op-amp characteristics (provided that the open-loop gain, A_o, is large relative to the closed-loop gain, A). Note from the graph of *Figure 1.4* that the signal bandwidth of such circuits equals the IC's f_T value divided by the circuit's closed-loop voltage gain value. Thus, the 741 gives a bandwidth of 100 kHz when the gain is set at × 10 (20 dB), or 1 kHz when the gain is set at × 1000 (60 dB).
  Figure 1.6(a) shows the op-amp wired in the closed-loop mode as a fixed-gain inverting ac amplifier. The voltage gain (A) is determined by the ratios of R_1 and R_2, and equals R_2/R_1, and the circuit's input impedance equals the R_1 value. This circuit can thus easily be designed to give any desired gain and input impedance values.

6 Basic principles and configurations

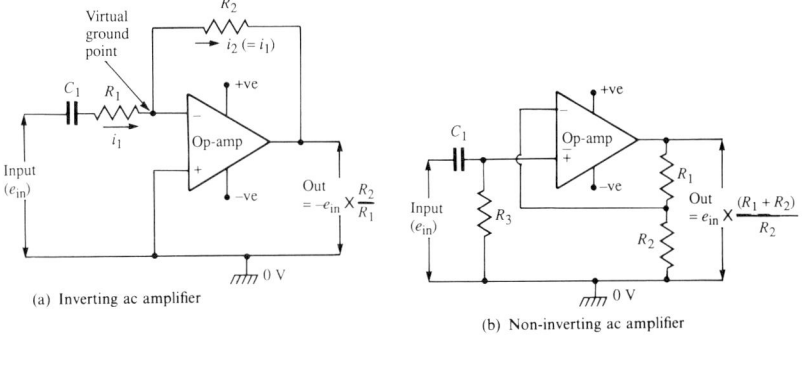

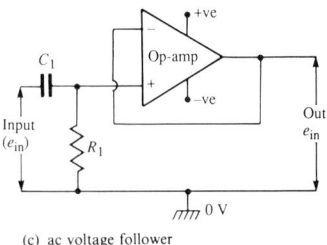

Figure 1.6 *Closed-loop ac amplifier circuits*

Note that although R_1 and R_2 control the gain of the complete circuit they have no effect on the voltage gain of the actual op-amp, and the signal voltage appearing at the output is thus A_o times greater than that appearing on the op-amp's input terminal, which thus acts as though it has an impedance of R_2/A_o connected between itself and ground. This terminal thus acts like a low impedance 'virtual ground' point. Also note that near-identical signal currents flow in R_1 and R_2, as is self-evident from the circuit's 'gain' formula.

Figure 1.6(b) shows how to connect the op-amp as a fixed-gain non-inverting ac amplifier. In this case the voltage gain equals $(R_1 + R_2)/R_2$, and the input impedance, looking into the input pin of the op-amp, approximately equals $(A_o/A)Z_{in}$, where Z_{in} is the open-loop input impedance of the op-amp. This impedance is shunted by R_3, however, so the input impedance of the actual circuit is less than the R_3 value.

The above circuit can be made to function as a precision ac voltage follower by wiring it in the unity-gain mode, as shown in *Figure 1.6(c)*, where the op-amp operates with 100% negative feedback. In this case the input impedance of the op-amp is very high (roughly $A_o \times Z_{in}$), but is shunted by R_1, which thus determines the circuit's input impedance value.

Op-amp parameters

Basic principles and configurations 7

An ideal op-amp would have infinite values of input impedance, gain and bandwidth, and have zero output impedance and give perfect tracking between input and output. Practical op-amps fall short of these ideals. Consequently, various performance parameters are detailed in op-amp data sheets, and indicate the 'goodness' of a particular device. The most important of these parameters are detailed below.

A_o *(open-loop voltage gain)*
This is the low-frequency voltage gain occurring between the input and output terminals of the op-amp, and may be expressed in direct terms or in terms of dB. Typical figures are × 100,000, or 100 dB.

Z_{in} *(input impedance)*
This is the resistive impedance looking directly into the input terminals of the op-amp when used open-loop. Typical values are 1 MΩ for op-amps with bipolar input stages, and a million MΩ for FET-input op-amps.

Z_o *(output impedance)*
This is the resistive output impedance of the basic op-amp when used open-loop. Values of a few hundred ohms are typical of most op-amps.

I_b *(input bias current)*
The input terminals of all op-amps sink or source finite currents when biased for linear operation. The magnitude of this current is denoted I_b, and is typically a fraction of a microamp in bipolar op-amps, and a few picoamps in FET types.

V_s *(supply voltage range)*
Op-amps are usually operated with split (positive and negative) supply rails, and these must be within maximum and minimum limits. If voltages are too high the op-amp may be damaged, and if too low the op-amp will not function correctly. Typical limits are ±3 V to ±15 V.

$V_{i(max)}$ *(input voltage range)*
Most op-amps will only operate correctly if their input terminal voltage is below the supply line values. Typically, $V_{i(max)}$ is one or two volts less than V_s.

V_{io} *(differential input offset voltage)*
In an ideal op-amp, perfect tracking would exist between the input and output terminals, and the output would register zero with both inputs grounded. In practice, slight imbalances within the op-amp cause the device to act as

8 Basic principles and configurations

though a small offset or bias voltage exists on its inputs under this condition. Typically, this 'differential input offset voltage' has a value of only a few millivolts, but when this voltage is amplified by the gain of the circuit in which the op-amp is used it may be sufficient to drive the op-amp output well away from the 'zero' value. Because of this, most op-amps have some facility for externally nulling out the effects of this offset voltage.

Common mode rejection ratio (CMRR)
An op-amp produces an output proportional to the difference between the signals on its two input terminals. Ideally, it should give zero output if identical signals are applied to both inputs simultaneously, i.e., in common mode. In practice, such signals do not entirely cancel out within the op-amp, and produce a small output signal. The ability of an op-amp to reject common mode signals is usually expressed in terms of 'common mode rejection ratio', i.e., the ratio of the op-amps gain with differential signals versus the gain with common mode signals. A CMRR value of 90 dB is typical of most op-amps.

f_T *(transition frequency)*
An op-amp typically gives a low-frequency voltage gain of about 100 dB, and in the interest of stability its open-loop frequency response is internally tailored so that the gain falls off as the frequency rises, and falls to unity at a transition frequency denoted f_T. Usually the response falls at a rate of 6 dB per octave (20 dB per decade). *Figure 1.4* shows the typical response curve of the type 741 op-amp, which has an f_T of 1 MHz and a low frequency gain of 106 dB.

Note that (as has already been pointed out) the bandwidth of a closed loop amplifier circuit depends on its gain value and on the f_T value of the op-amp, as indicated in *Figure 1.4*. Thus, a circuit with a bandwidth of 100 kHz at a gain of 20 dB will give a bandwidth of only 1 kHz at a gain of 60 dB; the f_T figure thus represents a gain-bandwidth product.

Slew rate
As well as being subject to normal bandwidth limitations, op-amps are also subject to a phenomenon known as output slew rate limiting, which has the practical effect of making the useful signal bandwidth inversely proportional to the magnitude of the output signal voltage of the op-amp. *Figure 1.7* shows the effect that slew rate limiting can have on the output of an op-amp that is fed with a square wave input. Slew rate is normally specified in terms of volts per microsecond, and values in the range 1 V/μs to 10 V/μs are usual with most types of op-amp.

Practical op-amps

Practical op-amps are available in a variety of types of IC construction (bipolar, MOSFET, JFET, etc.), and in a variety of types of packaging styles (plastic dual-in-line, metal-can TO5, etc.). Some of these packages house a single op-amp, and others house two or four op-amps, all sharing common supply line connections. *Figure 1.8* gives the parameter and outline details of

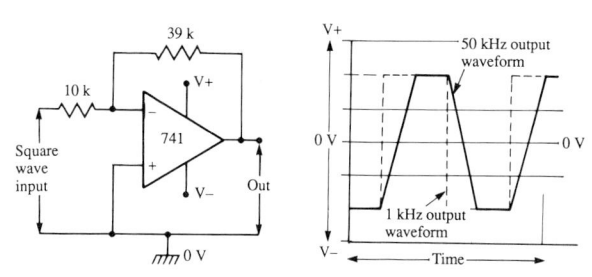

Figure 1.7 *Effects of slew-rate limiting on the output of an op-amp fed with a square wave input*

Parameter	Bipolar op-amps		MOSFET op-amps		JFET op-amps			
	741	NE531	CA3130E	CA3140E	LF351	LF411	LF441	LF13741
Supply voltage range	±3 V to ±18 V	±5 V to ±22 V	±2V5 to ±8 V OR 5 V to 16 V	±2 V to ±18 V OR 4 V to 36 V	±5 V to ±18 V			
Supply current	1.7 mA	5.5 mA	1.8 mA	3.6 mA	800 µA	1.8 mA	150 µA	2 mA
Input offset voltage	1 mV	2 mV	8 mV	5 mV	5 mV	0.8 mV	1 mV	5 mV
Input bias current	200 nA	400 nA	5 pA	10 pA	50 pA	50 pA	10 pA	50 pA
Input resistance	1M0	20M	1.5 TΩ	1.5 TΩ	1TΩ	1TΩ	1TΩ	0.5 TΩ
Voltage gain, A_o	106 dB	96 dB	110 dB	100 dB	88 dB	106 dB	100 dB	100 dB
CMRR	90 dB	100 dB	90 dB	90 dB	100 dB	100 dB	95 dB	90 dB
f_T	1 MHz	1 MHz	15 MHz	4.5 MHz	4 MHz	4 MHz	1 MHz	1 MHz
Slew rate	0.5 V/µS	35 V/µS	10 V/µS	9 V/µS	13 V/µS	15 V/µS	1 V/µS	0.5 V/µS
8-pin DIL outline	b	a	c	c	b	b	b	b

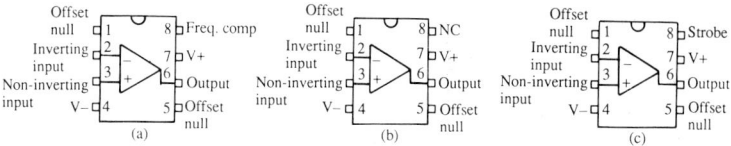

Figure 1.8 *Parameter and outline details of eight popular single op-amp types*

10 Basic principles and configurations

eight popular single op-amp types, all of which use 8-pin DIL (dual in-line) packaging, while *Figures 1.9* and *1.10* give details of popular dual and quad op-amp ICs respectively.

Parameter	Bipolar op-amp			MOSFET op-amps	JFET op-amps			
	LM358	NE5532	µA747C	CA3240E	LF353	LF442	TL072	TL082
Supply voltage range	±1.5 V to ±15 V	±3 V to ±20 V	±5 V to ±18 V	±2 V to ±18 V OR 4 V to 36 V	←±5 V to ±18 V→		←±2 V to ±18 V→	
Supply current	1.5 mA	8 mA	3 mA	8.4 mA	3.6 mA	0.4 mA	2.8 mA	2.8 mA
Input offset voltage	2 mV	0.5 mV	1 mV	5 mV	5 mV	1 mV	3 mV	5 mV
Input bias current	500 nA	200 nA	200 nA	10 pA	50 pA	10 pA	30 pA	30 pA
Input resistance	—	300 kΩ	1MΩ	1.5 TΩ	1TΩ	1TΩ	1TΩ	1TΩ
Voltage gain, A_o	100 dB	100 dB	104 dB	100 dB	100 dB	100 dB	106 dB	106 dB
CMRR	70 dB	100 dB	90 dB	90 dB	100 dB	95 dB	76 dB	76 dB
f_T	1 MHz	10 MHz	1 MHz	4.5 MHz	4 MHz	1 MHz	3 MHz	3 MHz
Slew rate	0.5 V/µS	9 V/µS	0.5 V/µS	9 V/µS	13 V/µS	1 V/µS	13 V/µS	13 V/µS
IC outline	a	a	b	a	a	a	a	a

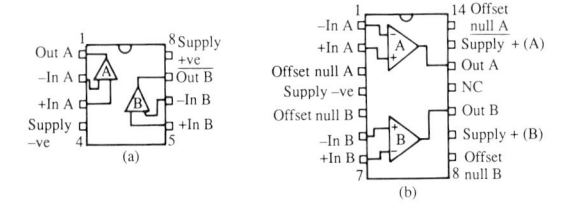

(a) Out A, –In A, +In A, Supply –ve / Supply +ve, Out B, –In B, +In B

(b) –In A, +In A, Offset null A, Supply –ve, Offset null B, –In B, +In B / Offset null A, Supply + (A), Out A, NC, Out B, Supply + (B), Offset null B

Figure 1.9 *Parameter and outline details of eight popular dual op-amp types*

Among the single op-amps shown in *Figure 1.8*, the 741 and NE531 are bipolar types. The 741 is a very popular general-purpose type featuring internal frequency compensation and full overload protection on inputs and output. The NE531 is a high-performance type with a very high output slew rate capability. An external compensation capacitor (100 pF), wired between pins 6 and 8, is needed for stability, but can be reduced to a very low value (1.8 pF) to give a very wide bandwidth at high gain.

The CA3130 and CA3140 are MOSFET-input op-amps that can operate from single or dual power supplies, can sense inputs down to the negative supply rail value, have very high input impedances (1.5 million MΩ), and have outputs that can be strobed. The CA3130 has a CMOS output stage; an external compensation capacitor (typically 47 pF) between pins 1 and 8

Parameter	Bipolar op-amps					JFET op-amps			
	LM324	3403	4136	LF347	LF444	TL064	TL074	TL084
Supply voltage range	±1.5 V to ±16 V OR 3 V to 32 V	±1.25 V to ±18 V OR 2.5 V to 36 V	±2.5 V to ±18 V	±5 V to ±18 V	±5 V to ±18 V	±2 V to ±18 V		
Supply current	1 mA	3 mA	7 mA	7.2 mA	0.8 mA	0.8 mA	5.6 mA	5.6 mA
Input offset voltage	2 mV	2 mV	0.5 mV	5 mV	3 mV	3 mV	3 mV	5 mV
Input bias current	45 nA	150 nA	40 nA	50 pA	10 pA	30 pA	30 pA	30 pA
Input resistance	—	—	5 MΩ	1 TΩ	1 TΩ	1 TΩ	1 TΩ	1 TΩ
Voltage gain, A_0	100 dB	100 dB	110 dB	100 dB	100 dB	75 dB	106 dB	106 dB
CMRR	70 dB	90 dB	100 dB	100 dB	95 dB	76 dB	76 dB	76 dB
f_T	1 MHz	1 MHz	3 MHz	4 MHz	1 MHz	1 MHz	3 MHz	3 MHz
Slew rate	0.5 V/μS	1.2 V/μS	1 V/μS	13 V/μS	1 V/μS	3.5 V/μS	13 V/μS	13 V/μS
IC outline	a	a	b	a	a	a	a	a

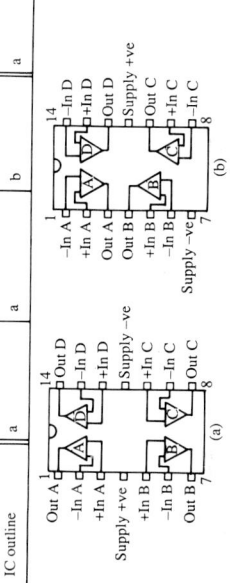

Figure 1.10 Parameter and outline details of eight popular quad op-amp types

12 Basic principles and configurations

permits adjustment of bandwidth characteristics. The CA3140 has a bipolar output stage and is internally compensated.

The LF351, 411, 441 and 13741 are JFET type op-amps with very high input impedances. The LF351 and 411 are high performance types, while the LF441 and 13741 are general-purpose types that can be used as direct replacements for the popular 741. Note that the LF441 quiescent current consumption is less than one tenth of that of the 741.

Looking next at the range of dual op-amps in *Figure 1.9*, the LM358, NE5532, and uA747C are all bipolar types. The LM358 can operate from single-ended power supplies, in which case its output can swing right down to zero volts; the NE5532 is a high-quality low-noise device; the uA747C is equal to two normal 741 op-amps sharing common power supply leads.

The CA3240E is a MOSFET device, and is simply a dual version of the very popular CA3140E op-amp.

The LF353, LF442, TL072 and TL082 range of devices all have JFET input stages. The LF353 is a low-cost device that can act as a replacement for a pair of 741 op-amps in most applications; the LF442 features a very low supply current consumption; and the TL072 and TL082 are low-cost high-performance types.

Finally, looking at the range of quad op-amps in *Figure 1.10*, the LM324 and the 3403 and 4136 are all bipolar devices, while all other devices have JFET input stages. The LM324 is a low-cost general-purpose unit, and the 3403 and 4136 are high performance devices. Among the JFET devices, the LF347, TL074 and TL084 are low-cost high-performance types, and the LF444 and TL064 feature low supply current consumption.

Offset nulling

Most single op-amps have an offset nulling facility, to enable the output to be set to precisely zero with zero input, and this is usually achieved by wiring a 10 k pot between pins 1 and 5 and connecting the pot slider (either directly or via a 4k7 range-limiting resistor) to the negative supply rail (pin 4), as shown in *Figure 1.11*. In the case of the CA3130, a 100 k offset nulling pot must be used.

Applications round up

Operational amplifiers are very versatile devices, and can be used in an almost infinite variety of linear amplifier and switching applications. *Figures 1.12* to *1.26* show a small selection of basic applications circuits that can be used, and which we shall be looking at in greater detail in the next few chapters. In most of these diagrams, the supply line connections have been omitted for clarity.

Basic principles and configurations 13

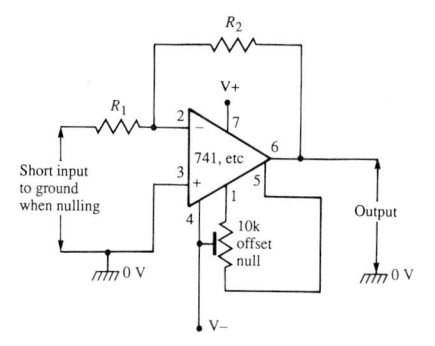

Figure 1.11 Typical offset nulling system

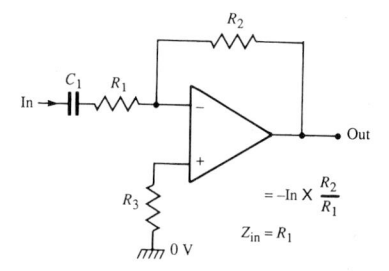

Figure 1.12 Inverting ac amplifier

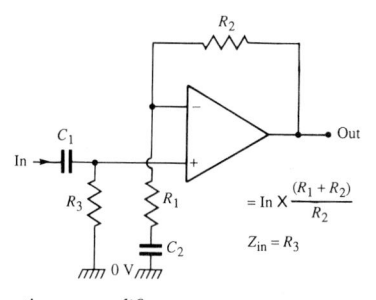

Figure 1.13 Non-inverting ac amplifier

Figures 1.12 and *1.13* show how op-amps can be used to make fixed-gain inverting and non-inverting ac amplifiers respectively. In both cases, the gain and the input impedance of the circuit can be precisely controlled by suitable component value selection.

Figure 1.14 shows how to make a differential or *difference* amplifier with a gain equal to R_2/R_1. If R_1 and R_2 have equal values, the circuit acts as an

14 Basic principles and configurations

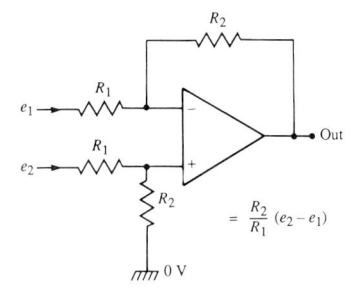

Figure 1.14 *Differential amplifier or analogue subtractor*

analogue subtractor. *Figure 1.15* shows the circuit of an inverting adder or audio mixer; if R_1 and R_2 have equal values, the inverted output is equal to the sum of the input voltages.

Op-amps can be made to act as precision active filters by wiring suitable R–C networks into their feedback loops. *Figures 1.16* and *1.17* show the basic connections for making second-order high-pass and low-pass filters respectively; these circuits have roll-offs of 12 dB/octave.

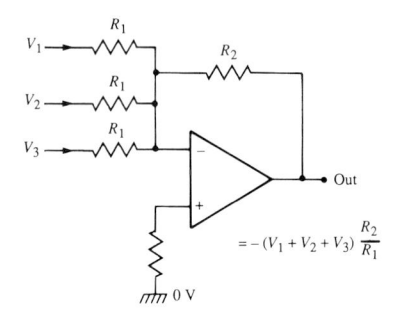

Figure 1.15 *Inverting analogue adder or audio mixer*

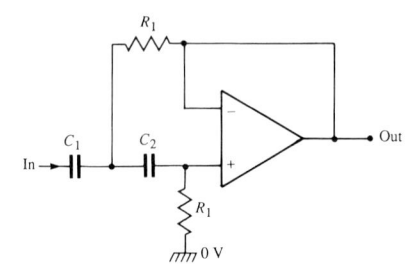

Figure 1.16 *High-pass second-order active filter*

Basic principles and configurations 15

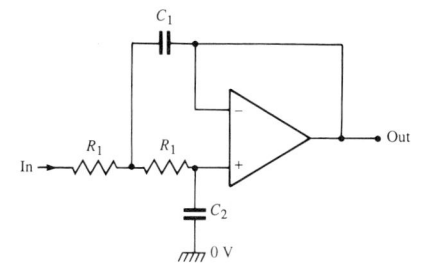

Figure 1.17 *Low-pass second-order active filter*

Figures 1.18 to *1.20* show some useful applications of the basic voltage follower or unity-gain non-inverting dc amplifier. The *Figure 1.18* circuit acts as a supply-line splitter, and is useful for generating split supplies from

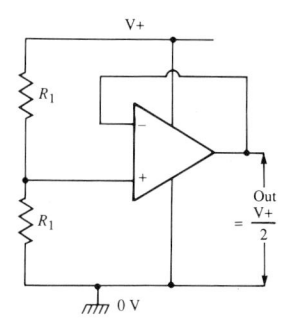

Figure 1.18 *Supply-line splitter*

single-ended ones. *Figure 1.19* acts as a semi-precision variable voltage reference, and *Figure 1.20* shows how the output current drive can be boosted so that the circuit acts as a variable voltage power supply.

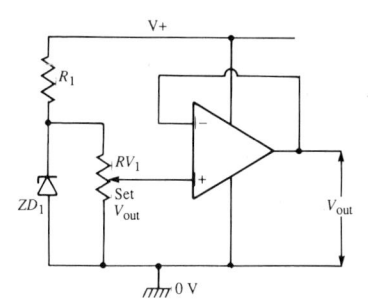

Figure 1.19 *Adjustable voltage reference*

16 Basic principles and configurations

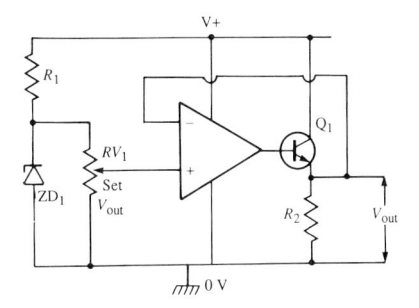

Figure 1.20 *Adjustable-voltage power supply*

Figure 1.21 shows the basic circuit of a bridge-balance detector, in which the output swings high when the inverting pin voltage is above that of the non-inverting pin, and vice versa. This circuit can be made to function as a precision opto- or thermo-switch by replacing one of the bridge resistors with a light-dependent resistor (LDR) or thermistor.

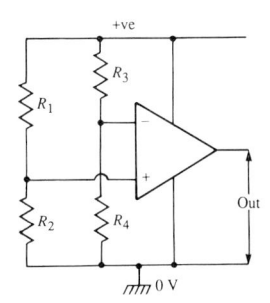

Figure 1.21 *Bridge-balance detector/switch*

Figures 1.22 to *1.23* show how to make precision half-wave rectifiers and ac/dc converters. These are very useful instrumentation circuits.

Finally, to complete this introductory chapter, *Figures 1.24* to *1.26* show some useful waveform generator circuits. The *Figure 1.24* design uses a Wien bridge network to generate a good sine wave; amplitude stabilization is obtained via a low-current lamp. *Figure 1.25* is a very useful square wave generator circuit in which the frequency can be controlled via any one of the passive component values. The frequency of the *Figure 1.26* function generator circuit can also be controlled via any one of its passive component values, but in this particular design generates both square and triangle waveforms.

Basic principles and configurations 17

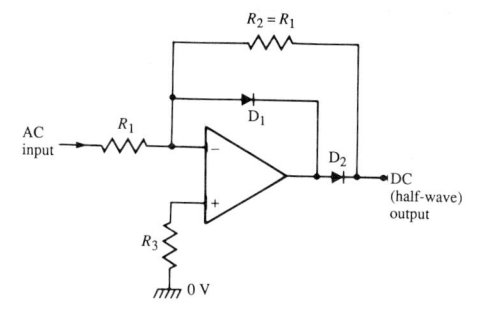

Figure 1.22 *Precision half-wave rectifier*

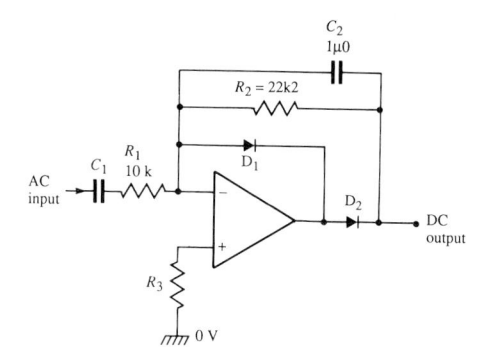

Figure 1.23 *Precision half-wave ac/dc converter*

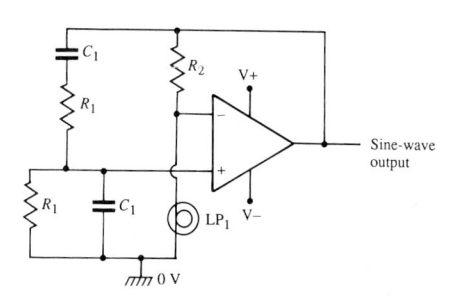

Figure 1.24 *Wien-bridge sine-wave generator*

18 Basic principles and configurations

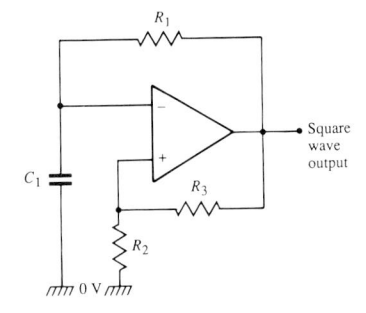

Figure 1.25 *Free-running multivibrator*

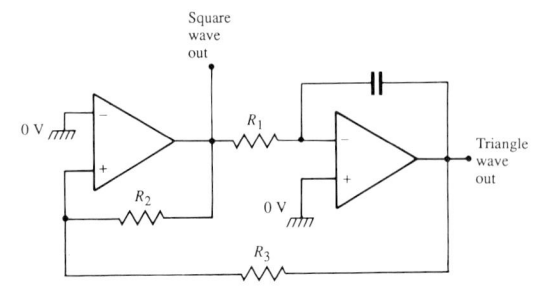

Figure 1.26 *Sine/square function generator*

2 Amplifiers and active filters

In Chapter 1 we took an in-depth look at the basic operating principles of conventional voltage-in voltage-out operational amplifiers, and showed some of the basic circuit configurations in which such op-amps can be used. In this chapter we concentrate on practical methods of using these op-amps as linear amplifiers and active filters.

When reading this chapter note that all circuits are shown designed around a standard 741 op-amp and operated from dual 9-V supplies, but that in practice these circuits will work with virtually any normal op-amp, and from any supply voltages within the operating range of the op-amp. If alternative op-amps are used, however, attention should (where applicable) be paid to possible differences in offset biasing networks.

Inverting amplifier circuits

Figure 2.1 shows the practical circuit of an inverting dc amplifier with an overall voltage gain (A) of $\times 10$, and with an offset nulling facility the output to be set to precisely zero with zero applied input. The voltage gain and input impedance are determined by the R_1 and R_2 values, and can be altered to suit individual user needs. The gain can be made variable by using a fixed and a variable resistor in place of R_2. For optimum biasing stability, R_3 should equal the parallel values of R_1 and R_2.

This circuit will work without the RV_1 offset nulling network facility, but its output will offset by an amount equal to the op-amp's input offset voltage value (typically 1 mV in a 741) multiplied by the circuit's closed-loop A value, e.g., if the circuit has a gain of $\times 100$, the output may offset by 100 mV with

20 Amplifiers and active filters

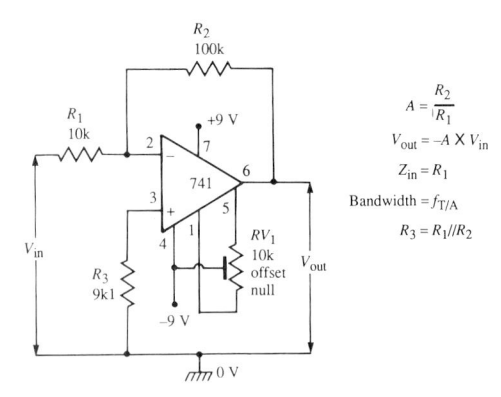

Figure 2.1 *Inverting dc amplifier with offset-nulling facility and × 10 voltage gain*

zero input applied. The circuit's bandwidth equals the f_T value (1 MHz in a 741) divided by the A value, e.g., it gives a bandwidth of 100 kHz with a gain of × 10, or 10 kHz with a gain of × 100.

The *Figure 2.1* circuit can be adapted for use as an ac amplifier by simply wiring a blocking capacitor in series with its input terminal, as shown in *Figure 2.2*. Note in this case that no offset nulling facility is needed, and that (for optimum biasing) R_3 is given a value equal to R_2.

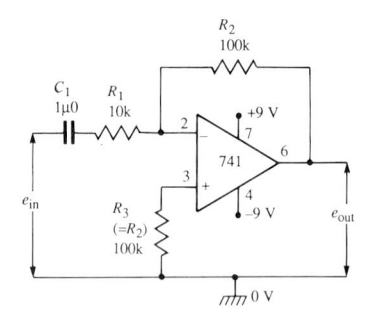

Figure 2.2 *Inverting ac amplifier with × 10 gain*

Non-inverting amplifier circuits

An op-amp can be used as a non-inverting dc amplifier with offset compensation by using the connections shown in *Figure 2.3*, which shows a × 10 amplifier. The voltage gain is determined by R_1–R_2 ratios, as indicated; if R_1 has a value of zero the gain falls to unity, but if R_2 has a value of zero the

Amplifiers and active filters   21

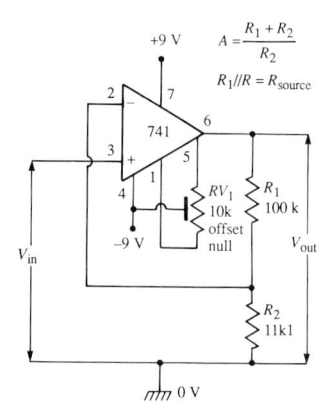

$$A = \frac{R_1 + R_2}{R_2}$$

$$R_1 // R = R_{source}$$

Figure 2.3 *Non-inverting dc amplifier with offset-nulling facility and ×10 gain*

gain rises to the open-loop gain value of the op-amp. The gain can thus be made variable by replacing R_2 with a pot and connecting its slider to the op-amp's inverting terminal, as shown in *Figure 2.4*, where the gain can be varied from ×1 to ×101 via RV_1.

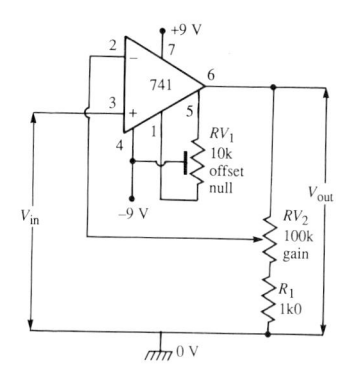

Figure 2.4 *Non-inverting variable-gain (×1 to ×101) dc amplifier*

Note that for correct operation the input (non-inverting) terminal of each of these circuits must have a dc path to the common (zero volts) rail. This path is normally provided via the dc input signal. In *Figure 2.3*, the parallel values of R_1 and R_2 should ideally (for optimum biasing) have a value equal to the input signal's source resistance.

A major feature of the non-inverting op-amp circuit is that it gives a very high input impedance. In theory, this impedance equals the open-loop input

22 Amplifiers and active filters

resistance (1 MΩ in a 741) multiplied by A_o/A. In practice, input impedance values of hundreds of megohms can easily be obtained in dc circuits such as those of *Figures 2.3* and *2.4*.

Figure 2.5 shows how the *Figure 2.3* circuit can be modified for use as a × 10 non-inverting ac amplifier by removing the offset biasing network, connecting the non-inverting terminal of the op-amp to ground via biasing resistor R_3, and connecting the input signal via a blocking capacitor. Note here that gain-control resistors R_1–R_2 are isolated from ground via C_2, so the op-amp's inverting terminal is subject to 100% dc negative feedback and the circuit has excellent dc stability, but C_2 has negligible impedance at practical operating frequencies and thus does not effect the voltage gain value. For optimum biasing, R_3 should have a value equal to that of R_1.

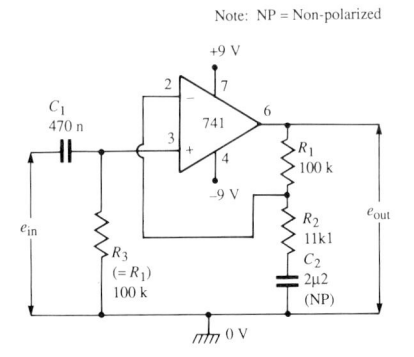

Figure 2.5 *Non-inverting × 10 ac amplifier with 100 k input impedance*

Clearly, the input impedance of the *Figure 2.5* circuit equals the R_3 value, and is limited to a maximum value of only a few megohms by practical considerations. *Figure 2.6* shows how this circuit can be further modified to give a very high input impedance (typically 50 MΩ). Here, the position of C_2 is moved relative to *Figure 2.5*, but this modification does not influence the gain or dc negative feedback characteristics of the circuit.

In the *Figure 2.6* circuit, however, the low end of R_3 is taken to ground via R_2, and the ac feedback signal appearing at the R_2–R_3 junction is virtually identical to that on the non-inverting input terminal of the op-amp. Near-identical signal voltages thus appear at both ends of R_3, which thus passes near-zero signal current. The apparent impedance of R_2 is thus increased to near-infinity by this 'bootstrap' action.

In practice, the input impedance of this circuit is typically limited to about 50 MΩ by leakage impedances of the actual op-amp socket and the printed circuit board (PCB) to which it is wired. Note that for optimum biasing the

Amplifiers and active filters 23

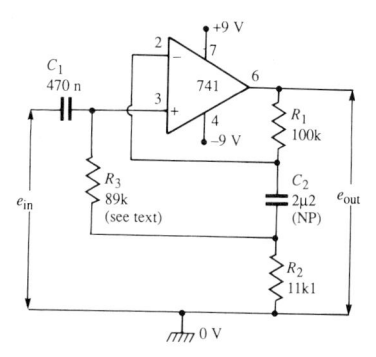

Figure 2.6 Non-inverting × 10 ac amplifier with 50 MΩ input impedance

sum of the R_2 and R_3 values should equal R_1, but in practice the R_3 value can differ from this ideal by up to 30%, and an actual value of 100 k can be used in the *Figure 2.6* circuit.

Voltage follower circuits

A voltage follower circuit produces an output voltage identical to that of the input signal, but has a very high-input impedance and a very low-output impedance. The circuit actually functions as a unity-gain non-inverting amplifier, with 100% negative feedback. *Figure 2.7* shows the 'idealized' design of a precision voltage follower with offset biasing. For optimum biasing, feedback resistor R_1 should have a value equal to the source resistance of the input signal.

In practice, the *Figure 2.7* circuit can often be greatly simplified. Eliminating the offset biasing network, for example, adds an error of only a

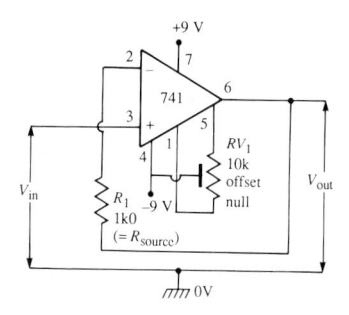

Figure 2.7 Precision dc voltage follower with offset null facility

24 Amplifiers and active filters

few millivolts to the output of the op-amp. Again, the value of feedback resistor R_1 can be varied over a wide range (from 0 to 100 k) without greatly influencing the output accuracy of the circuit. If an op-amp with a low f_T value (such as the 741) is used, the R_1 value can usually be reduced to zero. Note, however, that many high f_T op-amps tend towards instability when used in the unity-gain mode, and in such cases R_1 should be given a value of 1k0 or greater to effectively reduce the circuit bandwidth and thus enhance circuit stability.

Figure 2.8 shows an ac version of the voltage follower. Here, the input signal is dc blocked via C_1, and the op-amp's non-inverting pin is grounded via R_1, which determines the input impedance of the circuit. Ideally, feedback resistor R_2 should have the same value as R_1, but if R_2 has a high value it may significantly reduce the bandwidth of the circuit. This problem can be overcome by shunting R_2 with C_2, as shown dotted. If the latter technique is used with a high f_T op-amp, resistor R_3 can be connected as shown to ensure circuit stability.

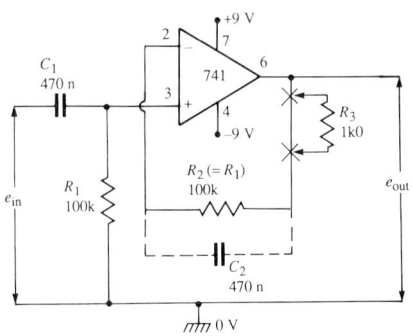

Figure 2.8 *ac voltage follower with 100 kΩ input impedance*

If a very high input impedance is needed from an ac voltage follower, it can be obtained by using the basic configuration shown in *Figure 2.9*, in which R_1 is bootstrapped from the op-amp output via C_2, so that the R_1 impedance is raised to near-infinity. In practice, this circuit will easily give an input impedance of 50 MΩ from a 741 op-amp, this limit being set by the leakage impedance of the op-amp's IC socket and the PCB.

If even greater input impedances are needed, the area of PCB surrounding the op-amp input pin should be provided with a printed 'guard ring' driven from the op-amp output, as shown, so that the PCB leakage impedances, etc., are themselves bootstrapped and raised to near-infinite values. In this case the *Figure 2.9* circuit gives an input impedance of about 500 MΩ when using a 741 op-amp, or even greater if a FET-input op-amp is used. *Figure 2.10* shows an example of the guard ring etched on a PCB.

Amplifiers and active filters 25

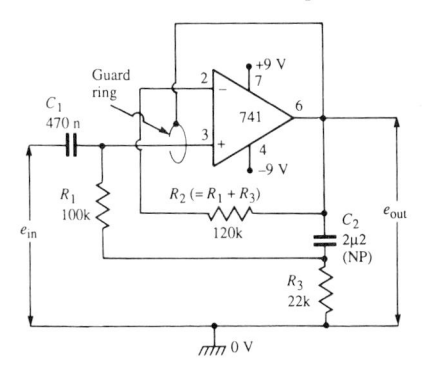

Figure 2.9 *ac voltage follower with 50 MΩ input impedance without the guard ring, or 500 MΩ with the guard ring*

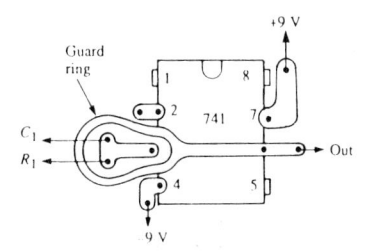

Figure 2.10 *Guard ring etched on a PCB and viewed through the top of the board*

Biasing accuracy

In the descriptions of the *Figure 2.1* to *2.9* circuits, great emphasis has been given to the selection of various component values to make them ideal for 'optimum biasing'. In practice, however, op-amps are very versatile devices and can accept considerable errors in the values of these components. *Figure 2.11* should help put the subject of 'biasing' into perspective.

Figure 2.11 shows the equivalent circuit of an idealized amplifier, in which the actual op-amp has zero intrinsic input offset voltage error, and the voltage gain (A) of the complete circuit is controlled by a negative feedback network. The op-amp is biased by wiring its input terminals to the ground (common) line via R_1 and R_2. The op-amp draws input bias currents, I_b, via these resistors, and thus generates a volt drop across each bias resistor.

For all practical purposes, the two input bias currents of an op-amp have virtually identical values. Consequently, if R_1 and R_2 have equal values, identical volt drops will occur across each resistor, thus giving zero

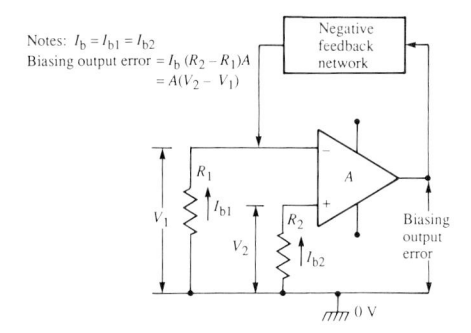

Figure 2.11 *Input biasing of an op-amp*

differential input voltage and zero biasing error at the output of the circuit; this is the 'ideal' biasing arrangement. If R_1 and R_2 do not have equal values, their volt drops will differ and will give an input differential error of $I_b (R_2 - R_1)$, and an output error that is A times greater than this value.

In practice, a bipolar op-amp such as the 741 has a typical I_b value of about 200 nA (0.2 μA), producing a volt drop of 0.2 mV across a 1k0 resistor. FET-input op-amps have typical I_b values of about 0.02 nA, giving a volt drop of a mere 0.02 μV across a 1 k 0 resistor. Thus, in *Figure 2.11*, a 741 op-amp will generate an input differential error of only 0.2 mV for each 1k0 difference in the R_1–R_2 values. If a FET-input op-amp is used, this error falls to a mere 0.02 μV. It can thus be seen that all of the *Figure 2.1* to *2.9* circuits can accept considerable latitude in their biasing component values. With this point in mind, let us look at some more amplifier circuits.

Current-booster follower circuits

Most op-amps can give maximum output currents of only a few milliamps, and this is the current-driving limit of the voltage follower circuits of *Figures 2.7* to *2.9*. The current-driving capacity of a voltage follower can easily be increased, however, by wiring a simple or a complementary emitter follower current booster stage between the op-amp output and the final output terminal of the circuit, as shown in the basic designs of *Figures 2.12* and *2.13*. Note that the base-emitter junctions of the transistors are wired into the negative feedback loop of the op-amp, to virtually eliminate the effects of junction non-linearity.

The *Figure 2.12* circuit can source large currents (via Q_1) but can sink only relatively small ones (via R_1); it can thus be regarded as a unidirectional

Amplifiers and active filters 27

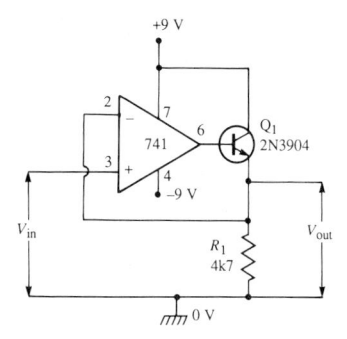

Figure 2.12 *Unidirectional dc voltage follower with boosted output current drive*

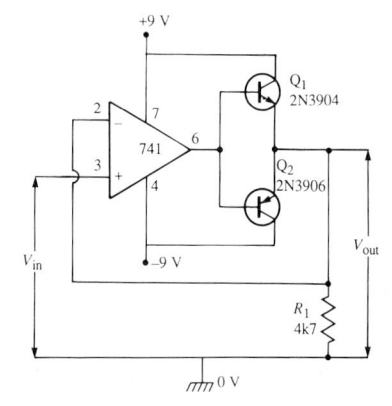

Figure 2.13 *Bidirectional dc voltage follower with boosted output current drive*

(positive-only) dc voltage follower. Several practical applications of this type of circuit are shown in Chapter 5.

The *Figure 2.13* circuit can both source (via Q_1) and sink (via Q_2) large output currents, and can thus be regarded as a bidirectional (positive and negative) voltage follower. In the simple form shown it produces significant crossover distortion as the output moves around the zero volts value. This distortion can be eliminated by suitably biasing Q_1 and Q_2, in which case the circuit can form the basis of a good hi-fi amplifier.

In practice, the *Figure 2.12* and *2.13* circuits have maximum current drive capacities of about 50 mA, this figure being dictated by the low power ratings of the specified transistors. Greater drive capacity can be obtained by using alternative transistors.

Adders and subtractors

Figure 2.14 shows the circuit of a unity-gain analogue dc voltage adder that gives an inverted output equal to the sum of the three input voltages. Input resistors R_1 to R_3 and feedback resistor R_4 have identical values, so the circuit acts as a unity-gain inverting dc amplifier between each input terminal and the output. The current flowing in R_4 is equal to the sum of the R_1 and R_3 currents, and the inverted output voltage is thus equal to the sum of the input voltages. In high-precision applications, the circuit can be fitted with an offset nulling facility.

The *Figure 2.14* circuit is shown with only three input connections, but can in fact be given any number of inputs (each with a value equal to R_1), but in this case the R_5 value should (for optimum biasing) be altered to equal the parallel values of all other resistors. If required, the circuit can be made to give a voltage gain greater than unity by simply increasing the R_4 value. It can be used as a multi-input audio mixer by ac-coupling the input signals and giving R_5 the same value as the feedback resistor, as shown in the 4-input circuit of *Figure 2.15*.

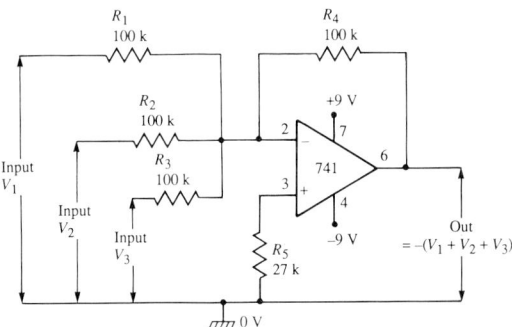

Figure 2.14 *Unity-gain inverting dc adder*

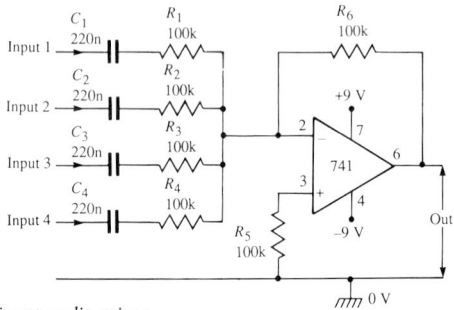

Figure 2.15 *4-input audio mixer*

Amplifiers and active filters   29

Figure 2.16 shows the circuit of a unity-gain dc differential amplifier or analogue subtractor, in which the output equals the difference between the two input signal voltages, i.e., it equals $e_2 - e_1$. In this type of circuit the component values are chosen such that $R_1/R_2 = R_3/R_4$, in which case the voltage gain, A, equals R_2/R_1. When, in *Figure 2.16*, R_1 and R_2 have equal values, the circuit gives unity overall gain, and thus acts as an analogue subtractor.

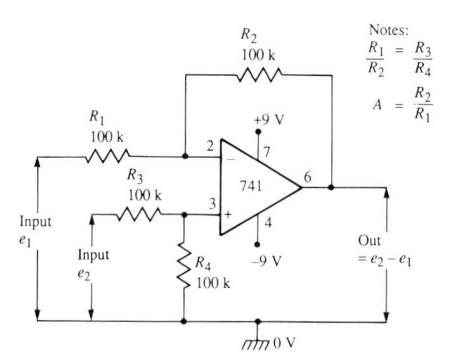

Figure 2.16 *Unity-gain dc differential amplifier, or subtractor*

Balanced phase splitter

A phase splitter has a pair of outputs which generate signals identical in amplitude and form but which are inverted (phase shifted by 180 degrees) relative to each other. *Figure 2.17* shows one way of making a balanced unity-gain dc phase splitter, using a pair of 741 op-amps. Here, IC_1 acts as a unity-gain non-inverting amplifier or voltage follower, and provides a buffered

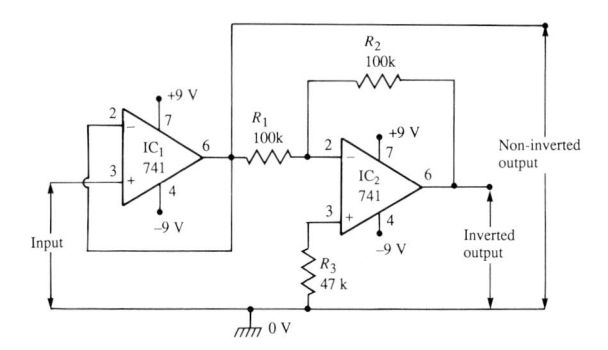

Figure 2.17 *Unity-gain balanced dc phase-splitter*

30   Amplifiers and active filters

output signal that is identical to that of the input. This output also provides the input drive to IC$_2$, which acts as a unity-gain inverting amplifier, and provides the second output, which is inverted but is otherwise identical to the original input signal.

Active filters

Filter circuits are used to reject unwanted frequencies and pass only those wanted by the designer. A simple R–C low-pass filter (*Figure 2.18(a)*) passes

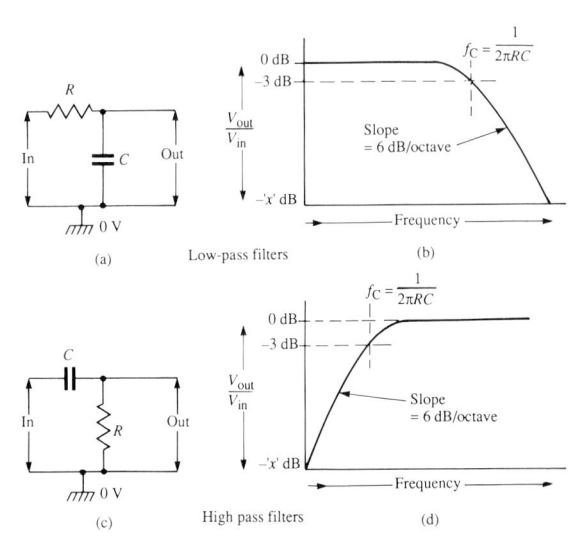

Figure 2.18   *Circuits and response curves of simple first-order* **R–C** *filters*

low-frequency signals but rejects high-frequency ones. The output falls by 3 dB at a 'break' or 'crossover' frequency (f_c) of $1/(2\pi RC)$, and then falls at a rate of 6 dB/octave (20 dB/decade) as the frequency is increased beyond this value (see *Figure 2.18(b)*). A 1 kHz low-pass filter thus gives about 12 dB of rejection to a 4 kHz signal, and 20 dB to a 10 kHz one, and so on.

A simple R–C high-pass filter (*Figure 2.18(c)*) passes high-frequency signals but rejects low-frequency ones. The output is 3 dB down at a break frequency of $1/(2\pi RC)$, and falls at a 6 dB/octave rate as the frequency is decreased below this value (see *Figure 2.18(d)*). Thus, a 1 kHz filter gives 12 dB of rejection to a 250 Hz signal, and 20 dB to 100 Hz, and so on.

Each of the above two filter circuits uses a single R–C or C–R stage, and is

Amplifiers and active filters 31

known as a first order filter. If we could simply cascade a number (n) of identical types of these filter stages, the resulting circuit would be known as an 'nth order' filter and would have an output slope, beyond f_c, of $6n$ dB/octave. Thus, a fourth order 1 kHz low-pass filter would have a slope of 24 dB/octave, and would give 48 dB of rejection to a 4 kHz signal, and 80 dB to a 10 kHz signal, and so on.

Unfortunately, simple R–C filters can not be directly cascaded, since they would then interact and give poor results; they can, however, be *effectively* cascaded by incorporating them into the feedback networks of suitable op-amp circuits. Such circuits are known as active filters, and *Figures 2.19* to *2.30* show practical examples of some of them.

Active filter circuits

Figure 2.19 shows the practical circuit and formula of a maximally-flat (Butterworth) unity-gain second-order low-pass filter with a 10 kHz break frequency. This circuit's output falls off at a rate of 12 dB/octave beyond 10 kHz, and is thus about 40 dB down at 100 kHz, and so on. To alter the break frequency, change either the R or the C value in proportion to the frequency ratio relative to *Figure 2.19*; reduce the values by this ratio to increase the frequency, or increase them to reduce the frequency. Thus, for 4 kHz operation, increase the R values by a ratio of 10 kHz/4 kHz, or 2.5 times.

$$f_c = \frac{1}{2.83\pi RC}$$

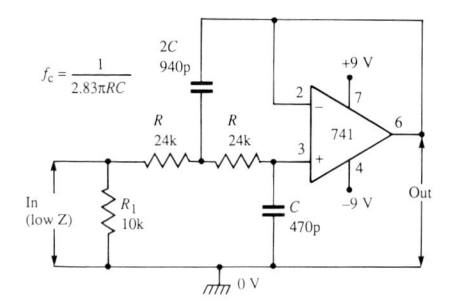

Figure 2.19 *Unity-gain second-order 10 kHz low-pass active filter*

A minor snag with the *Figure 2.19* circuit is that one of its C values should ideally be precisely twice the value of the other, and this can result in some rather odd component values. *Figure 2.20* shows an alternative second-order 10 kHz low-pass filter circuit that overcomes this snag and uses equal component values. Note here that the op-amp is designed to give a voltage gain of 4.1 dB via R_1 and R_2, which must have the values shown.

32 Amplifiers and active filters

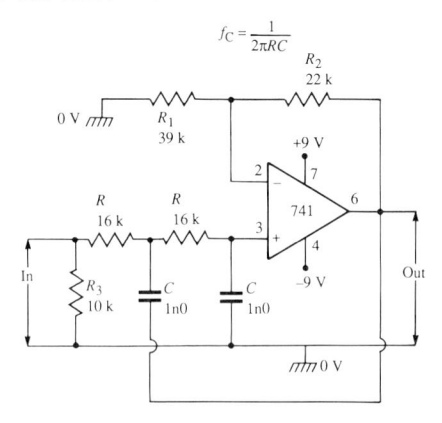

Figure 2.20 *Equal components version of second-order 10 kHz low-pass active filter*

Figure 2.21 shows how two of these equal component filters can be cascaded to make a fourth-order low-pass filter with a slope of 24 dB/octave. In this case gain-determining resistors R_1/R_2 have a ratio of 6.644, and R_3/R_4 have a ratio of 0.805, giving an overall voltage gain of 8.3 dB. The odd values of R_2 and R_4 can be made by series-connecting standard 5% resistors.

Figures 2.22 and *2.23* show unity-gain and equal component versions respectively of second-order 100 Hz high-pass filters, and *Figure 2.24* shows a fourth-order high-pass filter. The operating frequencies of these circuits, and those of *Figures 2.20* and *2.21*, can be altered in exactly the same way as in *Figure 2.19*, i.e., by increasing the R or C values to reduce the break frequency, or vice versa.

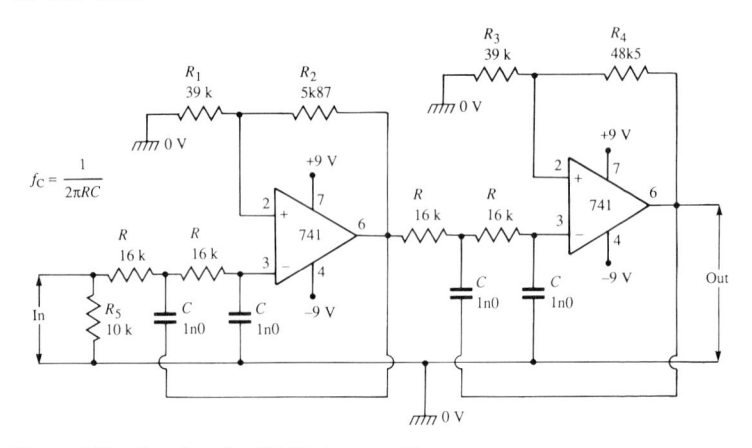

Figure 2.21 *Fourth-order 10 kHz low-pass filter*

Amplifiers and active filters   33

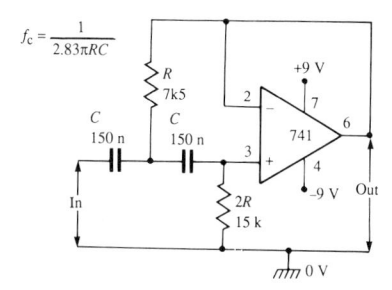

$$f_c = \frac{1}{2.83\pi RC}$$

Figure 2.22   *Unity-gain second-order 100 Hz high-pass filter*

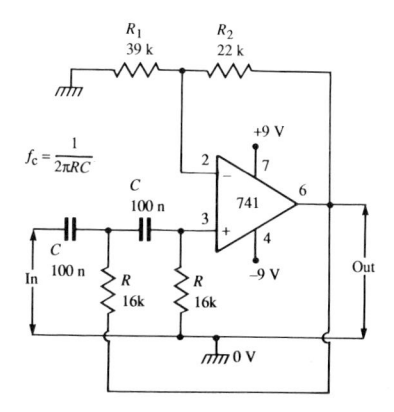

$$f_c = \frac{1}{2\pi RC}$$

Figure 2.23   *Equal components version of second-order 100 Hz high-pass filter*

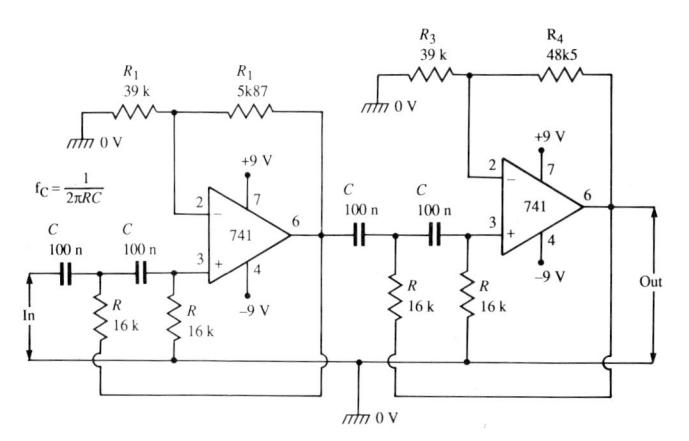

$$f_c = \frac{1}{2\pi RC}$$

Figure 2.24   *Fourth-order 100 Hz high-pass filter*

34 Amplifiers and active filters

Finally, *Figure 2.25* shows how the *Figure 2.23* high-pass and *Figure 2.20* low-pass filters can be wired in series to make (with suitable component value changes) a 300 Hz to 3.4 kHz 'speech' filter that gives 12 dB/octave of rejection to all signals outside of this range. In the case of the high-pass filter, the C values of *Figure 2.23* are reduced by a factor of three, to raise the break frequency from 100 Hz to 300 Hz, and in the case of the low-pass filter the R values of *Figure 2.20* are increased by a factor of 2.94 to reduce the break frequency from 10 kHz to 3.4 kHz.

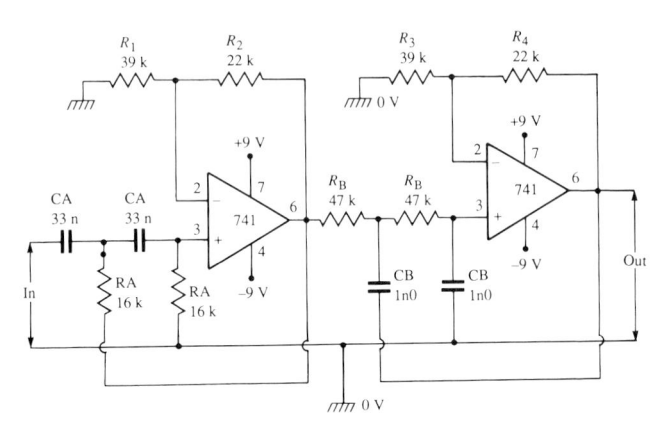

Figure 2.25 *300 Hz to 3.4 kHz speech filter with second-order response*

Variable active filters

The most useful type of active filter is that in which the crossover frequency is fully and easily variable over a fairly wide range, and *Figures 2.26* to *2.28* show three practical examples of second-order versions of such circuits.

The *Figure 2.26* circuit is a simple development of the high-pass filter of *Figure 2.22*, but has its crossover frequency fully variable fom 23.5 Hz to 700 Hz via RV_1. Note in this circuit that the resistive arms of the C–R networks have identical values (unlike *Figure 2.22*), so this design does not give maximally-flat Butterworth operation, but never the less gives a very good performance. This circuit can in fact be used as a high quality turntable disc (record) 'rumble' filter; 'fixed' versions of such filters usually have a 50 Hz crossover frequency.

The *Figure 2.27* circuit is a development of the high-pass filter of *Figure 2.19*, but has its crossover frequency fully variable from 2.2 kHz to 24 kHz via RV_1, and again does not give a maximally-flat Butterworth performance.

Amplifiers and active filters 35

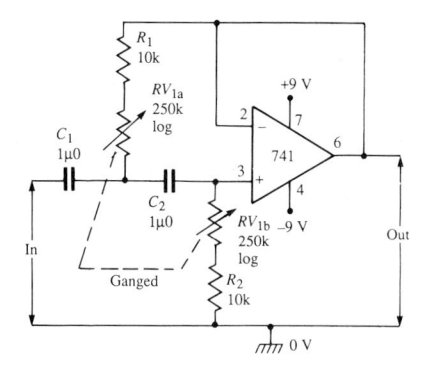

Figure 2.26 *Variable high-pass filter, covering 23.5 Hz to 700 Hz*

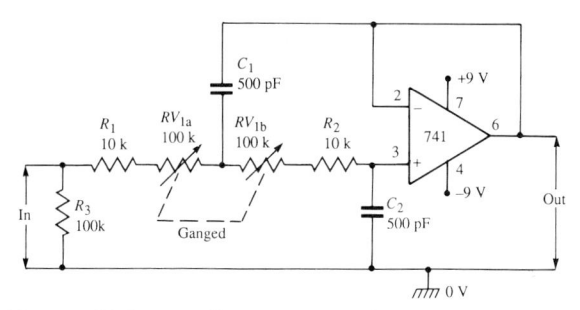

Figure 2.27 *Variable low-pass filter, covering 2.2 kHz to 24 kHz*

This circuit can in fact be used as a high quality scratch filter. Fixed versions of such filters usually have a 10 kHz crossover frequency.

Figure 2.28 shows how the above two filter circuits can be combined to make a really versatile variable high-pass/low-pass or rumble/scratch/speech filter. The high-pass crossover frequency is fully variable from 23.5 Hz to 700 Hz via RV_1, and the low-pass frequency is fully variable from 2.2 kHz to 24 kHz via RV_2.

Finally, to complete this chapter, *Figures 2.29* and *2.30* show two very useful active filter audio tone-control circuits, which enable the user to alter an audio system's frequency response to either suit his/her individual needs/moods, or to compensate for anomalies in room acoustics, etc. The *Figure 2.29* circuit is a conventional tone control design that can give up to 20 dB of boost or cut to bass or treble signals. The *Figure 2.30* circuit is similar, but has an additional filter control network that is centred on the 1 kHz 'midband' part of the spectrum, thus enabling this part of the audio band to also be boosted or cut by up to 20 dB.

36 Amplifiers and active filters

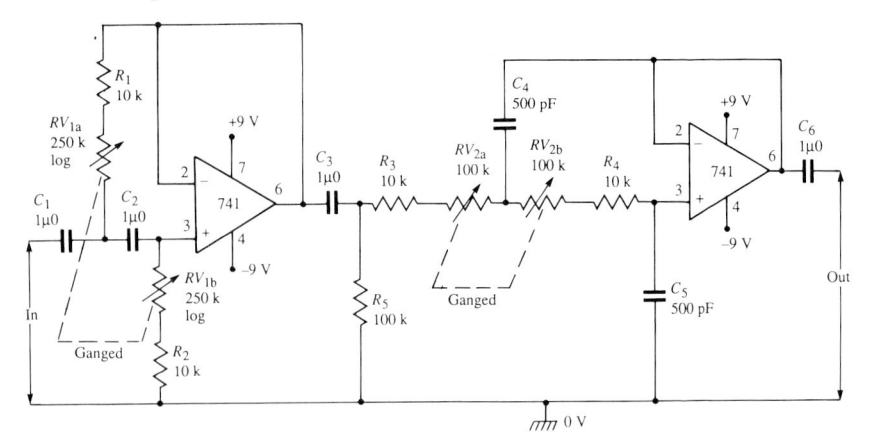

Figure 2.28 *Variable high-pass/low-pass or rumble/scratch/speech filter*

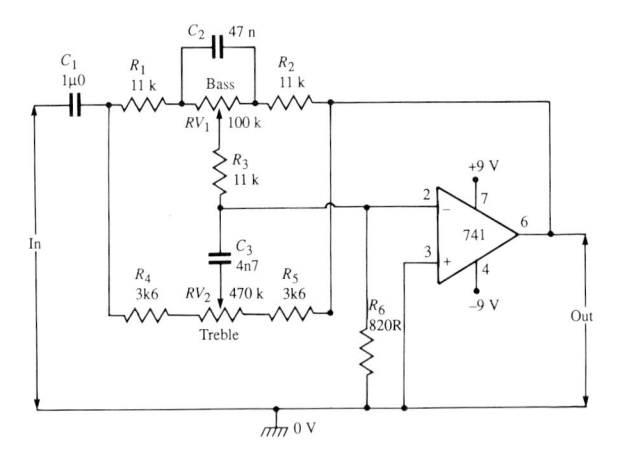

Figure 2.29 *Active tone control circuit*

Amplifiers and active filters 37

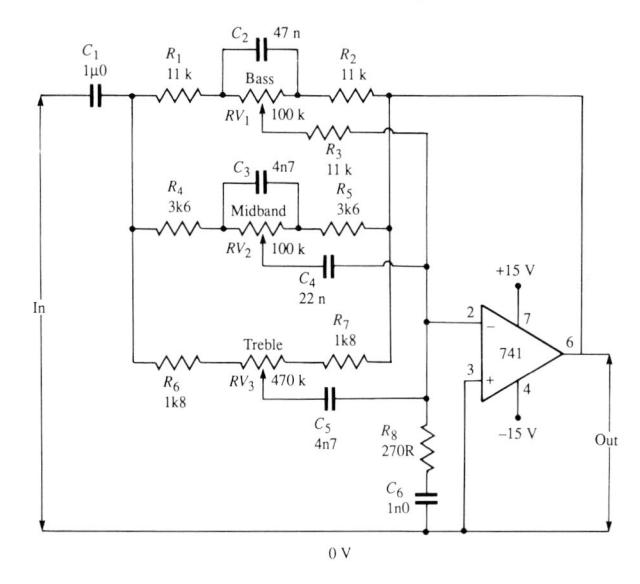

Figure 2.30 Three-band *(BASS, MIDBAND, TREBLE)* active tone control circuit

3 Voltage comparator circuits

A voltage comparator is a circuit that abruptly changes its output state when an input voltage, or a quantity that can be represented by a voltage (such as a current, resistance, temperature or light-level, etc.), goes beyond a pre-set reference value. Such circuits are usually designed around operational amplifiers.

Voltage comparators can be used as precision over-voltage or under-voltage switches, or can be made to activate relays or alarms, etc., when parameters such as load currents or temperatures or light levels go outside of, of come within, pre-set limits. They thus have many practical domestic and industrial uses, and a selection of these are described in this chapter.

Basic voltage comparator circuits

An easy way to make a voltage comparator is to use a CA3140 op-amp in one or other of the configurations shown in *Figures 3.1* and *3.2*. This particular op-amp is used because it can accept input voltages all the way down to its negative rail value, and its output can swing to within a couple of volts of its positive rail and to within a few millivolts of its negative supply rail value.

The operation of the *Figure 3.1* circuit is very simple. A fixed reference voltage (V_{ref}) is generated via R_2–ZD_1 and applied to the op-amp's non-inverting input terminal, and a test or input voltage is applied to its inverting terminal via current-limiting resistor R_1. When V_{in} is below V_{ref} the op-amp output is driven high (to positive saturation), but when V_{in} is above V_{ref} the output is driven low (to negative saturation), as shown in the diagram. Note that the CA3140 op-amp has a typical low-frequency open-loop voltage gain

Voltage comparator circuits 39

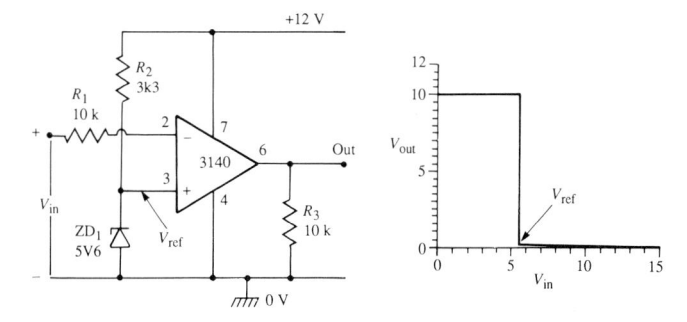

Figure 3.1 *Basic op-amp comparator that functions as an under-voltage switch; the output is high when V_{in} is below V_{ref}*

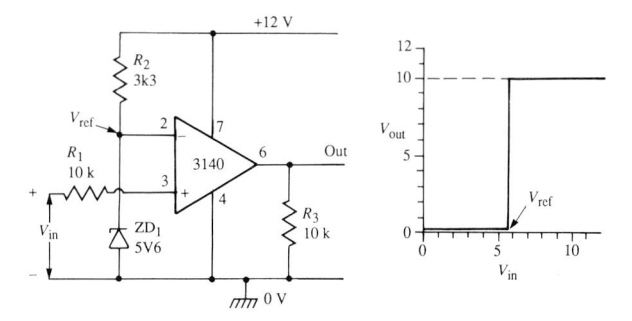

Figure 3.2 *Alternative op-amp voltage comparator that functions as an over-voltage switch; the output is high when V_{in} is above V_{ref}*

of 100 dB, so the circuit's output can be shifted from the high to the low state (or vice versa) by shifting the input voltage a mere 100 μV or so above or below the fixed reference voltage value.

The action of the above circuit can be reversed, so that its output is normally low but goes high when V_{in} exceeds V_{ref}, by simply transposing the op-amp's pin 2 and pin 3 connections, as shown in *Figure 3.2*.

Note in the above two voltage comparator circuits that V_{ref} can in fact have any value from zero to within 2 V of the positive supply rail value, so either circuit can be made to trigger at any desired value between these limits by simply interposing a pre-set pot between a fixed voltage reference source and the non-inverting terminal of the op-amp.

The inverting 'input' pin of the op-amp can also have any value from zero to within 2 V of the positive supply rail value. If the circuit is required to trigger at a high input voltage value this can be achieved by feeding the voltage to the op-amp input via a simple potential divider.

40 Voltage comparator circuits

Note that the above two voltage comparators give a non-regenerative switching action, in which the op-amp is driven into the linear (non-saturated) mode when the input voltage is within a few tens of microvolts of V_{ref}, and under this circumstance the op-amp output generates lots of spurious noise. In some applications this noise may be unacceptable, in which case the problem can be overcome by modifying the circuits so that a fraction of the op-amp's output voltage is fed back to the non-inverting input terminal, to give a regenerative switching action in which a degree of hysteresis or 'backlash' is imposed on the voltage switching levels.

Figures 3.3 to *3.7* show how the above-mentioned points can be put to practical use to make various types of 'special' voltage comparator circuits; many other variations are possible.

Special voltage comparator circuits

Figures 3.3 and *3.4* show how the basic comparator circuits can be modified to give variable voltage switching by using a pre-set pot (RV_1) to set the desired reference or trigger voltage at any value in the range 0 to 5V6, and to give regenerative (noiseless) switching by feeding part of the op-amp output back to the non-inverting terminal via R_3; note in the *Figure 3.4* circuit that the input is shunted via R_5, to ensure controlled hysteresis.

Figures 3.5 and *3.6* show examples of how the circuits can be modified to give high-value variable-voltage (0 to 130 V) triggering by interposing a simple potential divider (R_2–R_3) between the input signal and the input of the op-amp: The *Figure 3.5* circuit gives non-regenerative switching, while the *Figure 3.6* circuit gives regenerative switching.

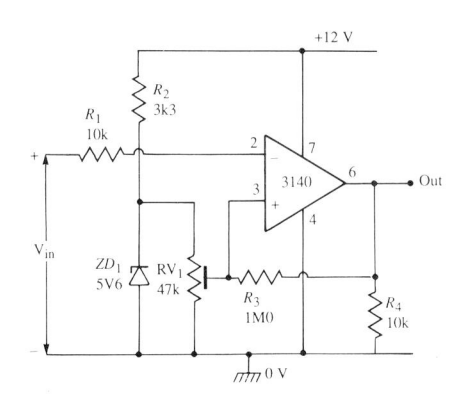

Figure 3.3 *Variable under-voltage switch with regenerative feedback*

Voltage comparator circuits 41

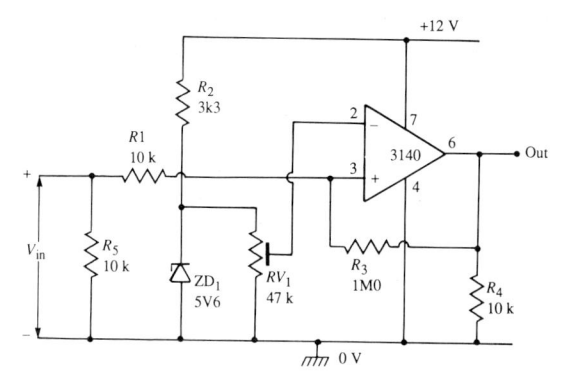

Figure 3.4 *Variable over-voltage switch with regenerative feedback*

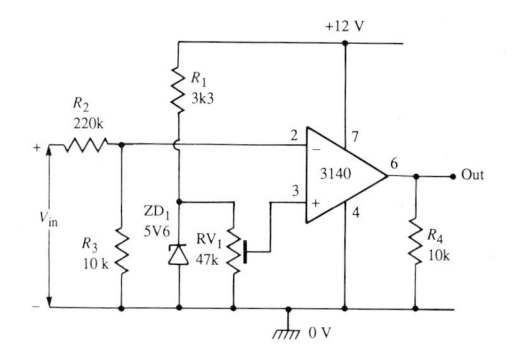

Figure 3.5 *High-value (0–130 V) under-voltage switch*

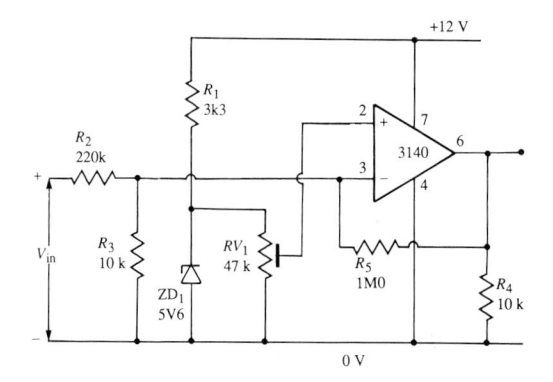

Figure 3.6 *High-value (0–130 V) regenerative over-voltage switch*

42 Voltage comparator circuits

Finally, *Figure 3.7* shows the comparator used as a sensitive audio sine-square converter that can operate from input signal amplitudes as low as 10 mV peak-to-peak at 1 kHz and which produces good square wave outputs from sine wave inputs with frequencies up to about 15 kHz. Input impedance is 100 k.

The operating theory of the *Figure 3.7* circuit is simple. Voltage divider R_1–R_2 and capacitor C_2 apply a decoupled reference voltage to pin 2 of the op-amp and an almost identical voltage is applied to signal input pin 3 via isolating resistor R_3. When a sine wave is fed to pin 3 via C_1 it swings pin 3 about the pin 2 reference level, causing the op-amp output to transition (switch) at the 'zero voltage difference' crossover points of the input waveform and produce a square wave output. Pre-set pot RV_1 is used to bias the op-amp so that its output is just pulled low with zero input signal applied, so that the circuit operates with maximum sensitivity and stability. Note that, because of the gain-bandwidth product characteristics of the op-amp, the circuit sensitivity decreases as the input frequency is increased.

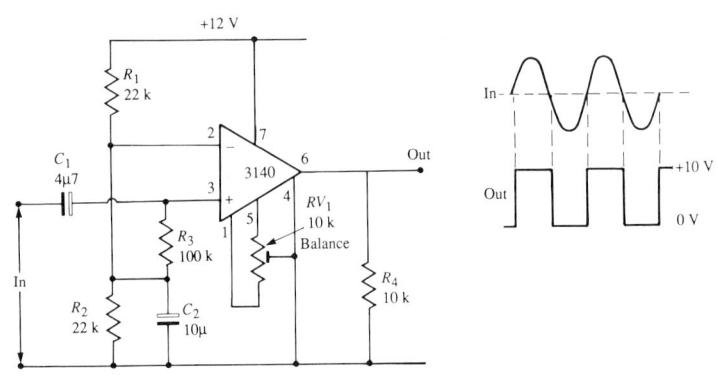

Figure 3.7 *This sensitive sine-square converter needs only a few tens of mV of input signal and produces a good square wave output up to about 15 kHz*

Window comparators

The voltage comparator circuits described so far give an output transition when their inputs go above or below a single reference voltage value. It is a fairly simple matter to interconnect a pair of voltage comparators so that an output transition is obtained when the inputs fall between, or go outside of, a pair of reference voltage levels. *Figure 3.8* shows the basic circuit configuration, which is generally known as a window comparator or discriminator.

The action of the *Figure 3.8* circuit is such that the output of the upper op-

Voltage comparator circuits 43

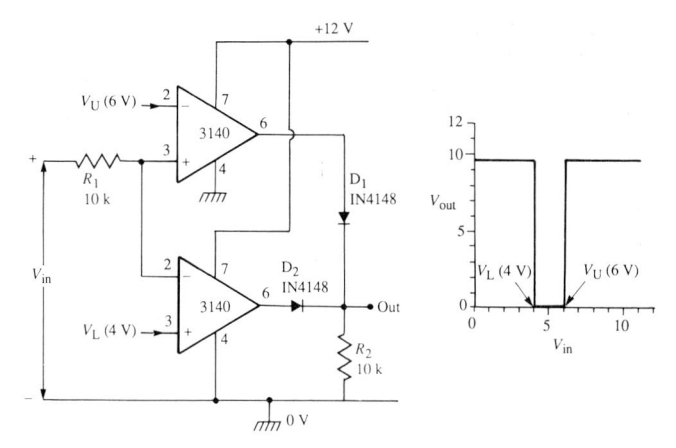

Figure 3.8 *A voltage window comparator or discriminator. The output goes high when V_{in} goes outside of the V_u or V_L limits*

amp goes high when V_{in} exceeds the 6 V V_u upper limit reference value, and the output of the lower op-amp goes high when V_{in} falls below 4 V V_L lower limit reference value. The outputs of the two op-amps are fed to R_2 via diode OR gate D_1–D_2, so the final output is low when V_{in} is within the limits set by V_u and V_L, but goes high whenever the input goes beyond these limits.

The action of the *Figure 3.8* circuit can be reversed, so that its output goes high only when the input voltage is within the window limits, by taking the output signal via a simple inverter stage. Alternatively, the required action can be obtained by transposing the two reference voltages and taking the output via a diode AND gate, as shown in *Figure 3.9*.

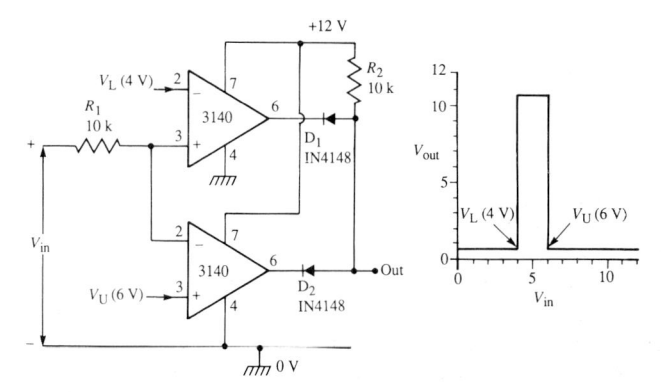

Figure 3.9 *An alternative window discriminator in which the output goes high when V_{in} falls within the V_u and V_L limits*

44 Voltage comparator circuits

Analogue-activated circuits

Voltage and windows comparators can easily be made to activate via any parameter that can be turned into an analogue voltage, and can thus be used to activate relays or alarms, etc., whenever temperatures, voltages, currents or light levels, etc., go outside of pre-set limits. *Figures 3.10* to *3.15* show some examples of practical analogue activated comparator circuits.

Figure 3.10 shows how a comparator can be made to function as an over-current switch that gives a high output when the load current exceeds a value pre-set via RV_1; the value of R_x is chosen so that it develops about 100 mV at the required trip current level. Thus, a fixed half-supply reference voltage is fed to pin 3 of the op-amp via R_3–R_4 and a similar but current-dependent voltage is fed to pin 2 via R_x–R_1–RV_1–R_2. In effect, these two sets of components are configured as a Wheatstone bridge, with one side feeding pin 3 and the other side feeding pin 2, and the op-amp is used as a bridge-balance detector. Consequently, the trip points of the circuit are not significantly influenced by supply voltage variations but are highly sensitive to load current variations.

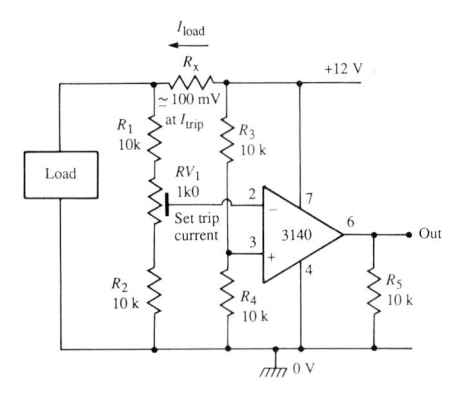

Figure 3.10 *An over-current switch; the output goes high when the load current exceeds a pre-set value*

Note that the action of the *Figure 3.10* circuit can be reversed, so that it functions as an under-current switch, by simply transposing the connections to pins 2 and 3 of the op-amp. The circuit can then be used as a lamp failure or load failure indicator in cars or in test gear, etc.

Figure 3.11 shows a sensitive ac over-voltage switch which gives a high output when the input signal exceeds a peak value (6 mV to 111 mV) pre-set via RV_1. The ac input signal is fed to the input of non-inverting variable-gain amplifier IC_1, which has its gain variable from × 45 to × 850 via RV_1 and has

Voltage comparator circuits 45

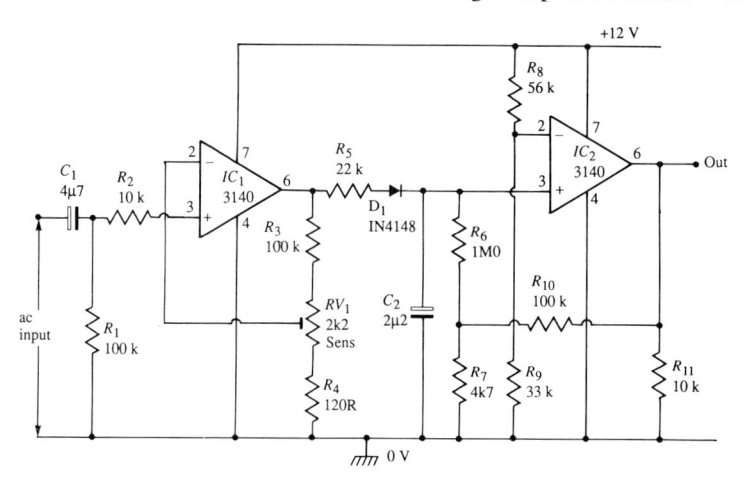

Figure 3.11 *This ac over-voltage switch can be triggered by input signals in the range 6 mV to 111 mV peak*

its input dc grounded via R_1–R_2 so the op-amp responds only to the positive half-cycles of the input signal. Thus, the output of IC_1 is an amplified but positively half-wave rectified version of the input signal; this signal is peak-detected via R_5–D_1–C_2–R_6–R_7 and fed to the input of non-inverting voltage comparator IC_2, which thus gives a positive output when the C_2 voltage exceeds the value on the junction of R_6–R_9.

Figures 3.12 to *3.15* show a variety of ways of using comparator circuits as light- or temperature-activated switches. All of these circuits use a light- or temperature-sensitive transducer (an LDR or cadmium sulphide photocell for light, or a negative-temperature-coefficient thermistor for temperature) as the sensing element and use the element as one arm of a Wheatstone bridge and use the op-amp as a simple bridge-balance detector, so that the trip point of each circuit is independent of supply line variations. In all cases, the sensing element must have a resistance in the range 5k0 to 100 k at the required 'trip' point and RV_1 is chosen to have the same resistance value as the sensing element at the required trip level.

The *Figure 3.12* to *3.15* circuits also show a variety of ways of using the output of the op-amp to activate a relay or generate an audible alarm signal. Thus, the *Figure 3.12* over-temperature switch has a transistor-driven relay output, while the *Figure 3.13* under-temperature switch has a VFET-driven relay output. Similarly, the light-operated switch circuit of *Figure 3.14* generates a monotone alarm output signal in a small speaker, while the dark-operated switch of *Figure 3.15* generates a low-power pulsed-tone signal in a small acoustic transducer.

46 Voltage comparator circuits

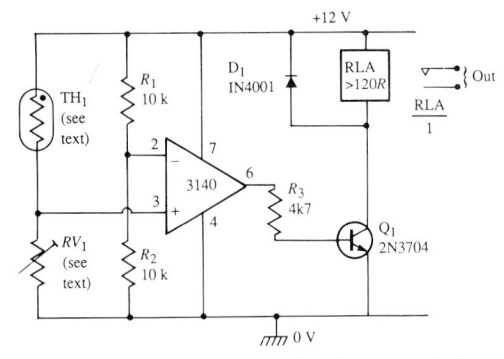

Figure 3.12 *Precision over-temperature switch with transistor/relay output*

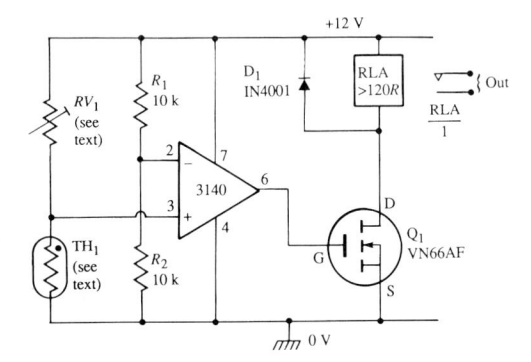

Figure 3.13 *Precision under-temperature switch with VFET/relay output*

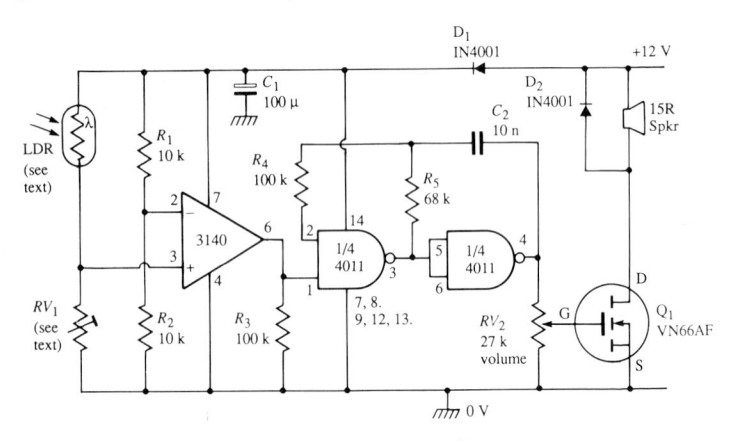

Figure 3.14 *Light-operated switch with monotone alarm output*

Voltage comparator circuits 47

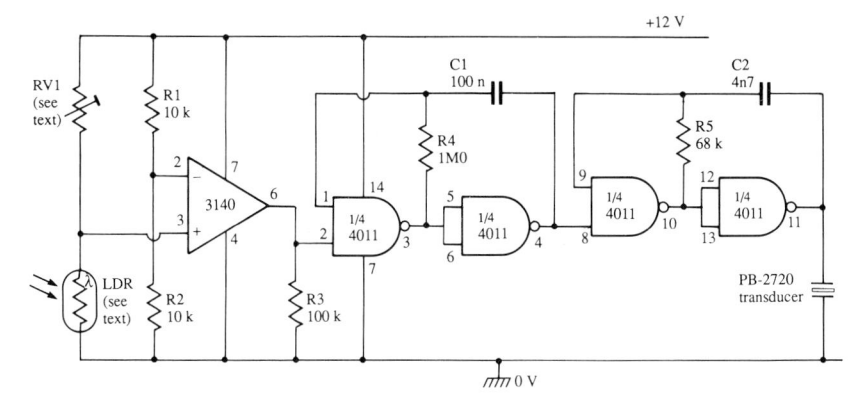

Figure 3.15 Dark-operated switch with low-power pulsed-tone output

Micro-power operation

All the 3140-based comparator circuits that we have looked at so far are continuously powered; they draw running currents of about 4 mA per op-amp and will thus drain a PP9 (or similar) supply battery in less than two days of continuous operation; they are thus not well suited to battery operation in 'portable' applications. In practice, however, all of these circuits can easily be modified for long-life battery operation by using a micro-power 'sampling' technique; the principle can be explained simply, as follows.

The *Figure 3.13* under-temperature switch circuit monitors temperature continuously and draws about 5 mA of quiescent current with the relay off. In reality, however, temperature is a slowly varying parameter and thus does not need to be monitored continuously; instead, it can be efficiently monitored by briefly 'inspecting' or 'sampling' it (by connecting the supply power and inspecting the op-amp output) only once every second or so.

Thus, if the sample periods are very brief (say, 300 μs) relative to the sampling interval (1 s), the MEAN current consumption of the monitor can be reduced by a factor equal to the interval/period ratio (e.g., by a factor of 3300) by using the sampling technique, so that, for example, the 5 mA consumption of the *Figure 3.13* circuit can be reduced to a MEAN value of a mere 1.6 μA, thus giving years of continuous operation from a PP9 battery. The 'sampling' technique thus enables true micro-power monitor or comparator designs to be implemented.

Figure 3.16 shows the basic circuit of a micropower or sampling version of the *Figure 3.13* under-temperature switch, which operates the relay when the TH$_1$ temperature falls below a pre-set value but which draws a mean

48 Voltage comparator circuits

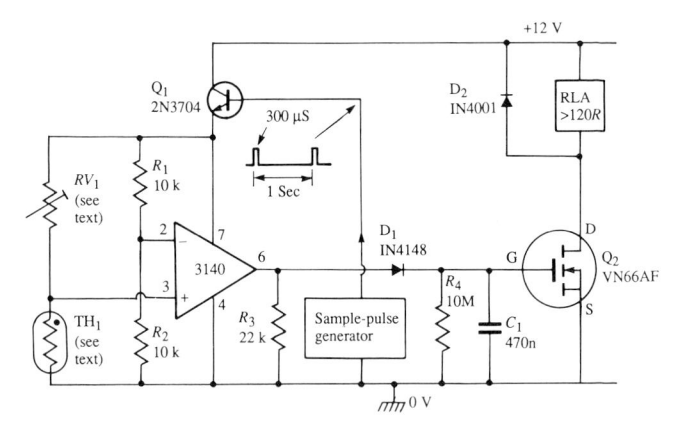

Figure 3.16 *This micro-power or 'sampling' version of the* Figure 3.13 *under-temperature switch draws a mean quiescent current of only a few microamps*

quiescent current of only a few microamps. The TH_1–RV_1–R_1–R_2–IC_1 monitor network is almost identical to that of *Figure 3.13*, but instead of being continuously powered it is powered via a 300 μs pulse just once every second via a sample-pulse generator and Q_1. Note that the op-amp output is fed to temporary memory store R_4–C_1 via D_1, and that the memory store operates the relay via VFET Q_2.

Thus, if the TH_1 temperature is outside of the trip level when the sample pulse arrives the op-amp output will remain low and no charge will be fed to C_1, so Q_2 and the relay will be off, but if the TH_1 temperature is within the trip level when the sample pulse arrives the op-amp output will switch high for the duration of the pulse and thus rapidly charge C_1 up via D_1 and thence drive the relay on via Q_2; the C_1 charge will then easily hold the relay on until the arrival of the next sample pulse.

The *Figure 3.16* circuit, then, illustrates the basic principles of the micro-power sampling technique. In reality the sampling intervals and pulse width used (and thus the reduction in mean power consumption) will depend on the specific application. If, for example, it is necessary to monitor transient changes in light or sound levels and these transients have minimum durations of 100 ms, this can be done by using a 50 ms sampling interval and, say, a 500 μs sample pulse, in which case the mean consumption of the circuit will be reduced by a factor of 100.

In some cases it may be necessary to slightly modify the operating principle of the sampling circuitry to obtain the desired micro-power operation. *Figure 3.17*, for example, shows how the principle may be adapted to make a coded-light-beam detector, in which the 'code' light signal is modulated at 1 kHz for

Voltage comparator circuits 49

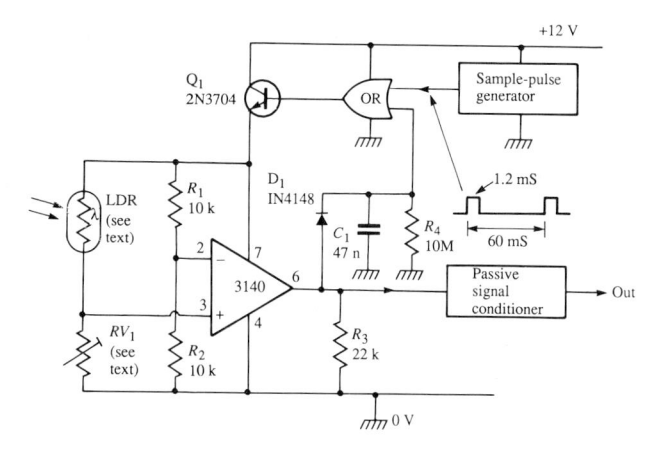

Figure 3.17 *This coded light beam detector circuit uses a modified version of the micropower 'sampling' technique*

a minimum duration of 100 ms. Thus, the sample-pulse generator is designed to produce a minimum pulse width of 1.2 ms so that it can capture at least one full 1 kHz code cycle, and the sampling interval is set at 60 ms so that part of a tone burst will always be captured. The sampling circuitry thus gives a 50:1 reduction in monitor current consumption.

Thus, in the *Figure 3.17* circuit, the sample generator repeatedly feeds 1.2 ms inspection pulses to the 3140 detector circuitry via one input of the OR gate and via Q_1 to see if any trace of a coded signal exists. If no trace of a code signal is detected the output of the op-amp remains low and another sample pulse is applied 60 ms later, but if a trace of a code signal is detected the output of the op-amp immediately switches high and the resulting pulse is captured by C_1 via D_1 and applied to the remaining input of the OR gate, thereby temporarily applying full power to the 3140 circuitry so that the code signal can be properly inspected via the passive signal conditioning circuitry to see if it conforms to the specified 'code' characteristics.

A sample-pulse generator

Note that for a sampling system to be truly efficient, its built-in sample-pulse generator must itself consume a total mean current of only a few microamps, and must thus be of fairly unusual design. The best way to make such a generator is to design it around the 4007UB CMOS IC, which houses two pairs of complementary MOSFETs, plus a simple CMOS inverter stage. *Figure 3.18* shows the functional diagram of this device.

50 Voltage comparator circuits

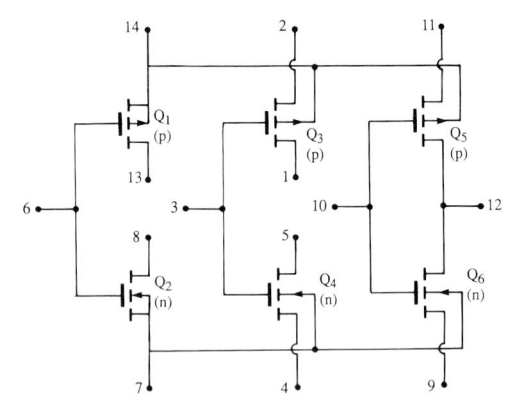

Figure 3.18 *Functional diagram of the 4008UB dual CMOS pair plus inverter*

A major attraction of the 4007UB IC is that its individual elements can easily be accessed and configured to make any of a variety of digital inverters or gates, or micropower linear amplifiers, oscillators or multivibrators. *Figure 3.19* shows how it can be configured to make a dual-time-constant astable multivibrator that can be used as a practical micro-power sample-pulse generator.

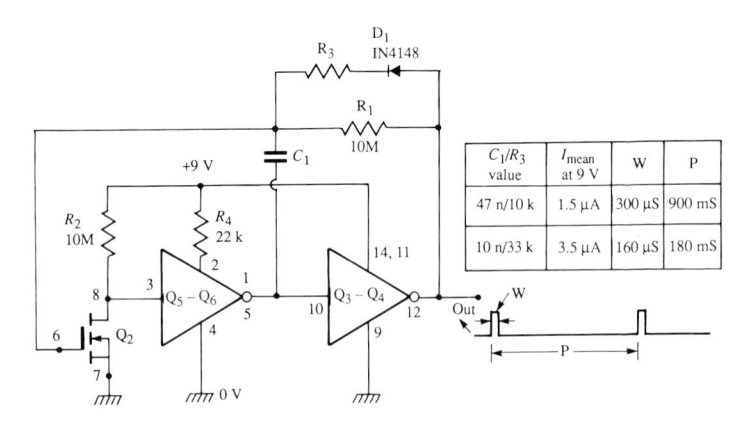

C_1/R_3 value	I_{mean} at 9 V	W	P
47 n/10 k	1.5 µA	300 µS	900 mS
10 n/33 k	3.5 µA	160 µS	180 mS

Figure 3.19 *Practical micro-power sample-pulse generator circuit*

In the *Figure 3.19* astable, timing capacitor C_1 alternately charges up via low-value resistor R_3 and discharges via high-value resistor R_1, to give the mark-space ratio waveforms shown. When C_1 and R_3 have values of 47 nF and 10 kΩ, the circuit generates a 300 µs pulse at intervals of 900 ms, and consumes a mean current of only 1.5 μA from a 9 V supply. When the C_1 and R_3 values are 10 nF and 33 kΩ, the pulse width is 160 μs and the interval is 180 ms, and the mean current is 3.5 μA. Other width and interval values can easily be obtained, to satisfy individual sample-pulse generator requirements, by simply changing the C_1 and R_3 values.

4 Waveform generator circuits

In the previous three chapters we have taken in-depth looks at the operating principles of conventional voltage-in voltage-out operational amplifiers, and at a variety of ways of using them to make practical linear amplifiers, active filters, and voltage comparators. In this chapter we look at ways of using these op-amps to make a variety of waveform-generating oscillators and switching circuits.

Sine wave oscillators

An op-amp can be used as a sine wave oscillator by wiring it in the basic configuration shown in *Figure 4.1*, in which the amplifier output is fed back to its input via a frequency-selective network, and the overall gain of the amplifier is controlled via a level-sensing system. For optimum sine wave generation, the feedback network must give an overall phase shift of zero degrees and a gain of unity at the desired frequency. If the gain is less than unity the circuit will not oscillate, and if greater than unity the output waveform will be distorted.

One practical way of implementing the above principle is to connect a Wien bridge network and an op-amp in the basic configuration shown in *Figure 4.2*. The frequency-sensitive Wien network is constructed from R_1–C_1 and R_2–C_2; normally, the network is symmetrical, so that $C_1 = C_2 = C$, and $R_1 = R_2 = R$. The main feature of the Wien network is that the phase relationship of its output and input signals varies from $-90°$ to $+90°$, and is precisely $0°$ at a centre frequency (f_o) of $1/2\pi CR$, or $1/6.28CR$. At this centre frequency the network has a voltage gain of $\times 0.33$.

Waveform generator circuits 53

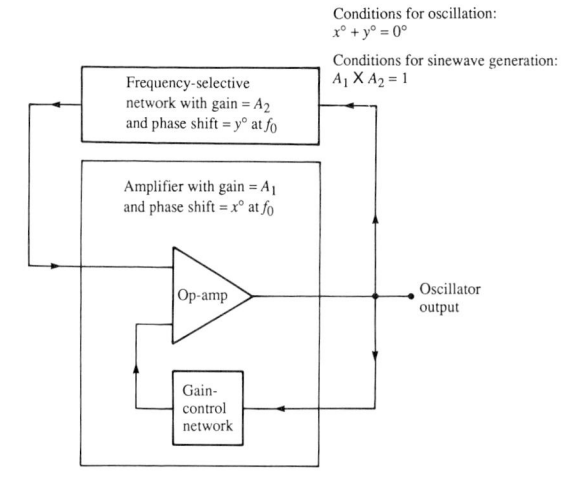

Figure 4.1 *Conditions for stable sine wave oscillation*

Conditions for oscillation:
$x° + y° = 0°$

Conditions for sinewave generation:
$A_1 \times A_2 = 1$

Frequency-selective network with gain = A_2 and phase shift = $y°$ at f_0

Amplifier with gain = A_1 and phase shift = $x°$ at f_0

Op-amp

Gain-control network

Oscillator output

Thus, in *Figure 4.2*, the Wien network is connected between the output and the non-inverting input of the op-amp, so that the circuit gives zero overall phase shift at f_0, and the actual amplifier is given a voltage gain of $\times 3$ via feedback network R_3–R_4, to give the total system an overall gain of unity. The circuit thus provides the basic requirements for sine wave oscillation. In practice, however, the ratios of R_3–R_4 must be carefully adjusted to give the overall voltage gain of precisely unity that is necessary for low-distortion sine wave generation.

The above circuit can easily be modified to give automatic gain adjustment and amplitude stability by replacing the passive R_3–R_4 gain-determining network with an active gain-control network that is sensitive to the amplitude

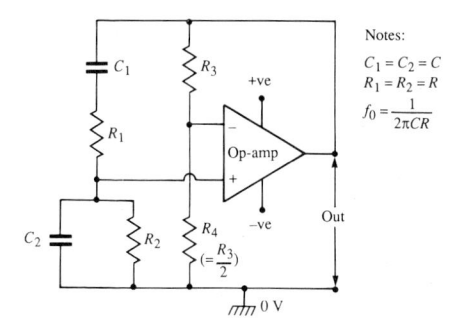

Notes:
$C_1 = C_2 = C$
$R_1 = R_2 = R$
$f_0 = \dfrac{1}{2\pi CR}$

Figure 4.2 *Basic Wien bridge sine wave oscillator*

54 Waveform generator circuits

of the output signal, so that gain decreases as the mean output amplitude increases, and vice versa. *Figures 4.3* to *4.7* show some practical versions of Wien bridge oscillators with automatic amplitude stabilization.

Thermistor stabilization

In the 1 kHz fixed-frequency oscillator circuit of *Figure 4.3* the output amplitude is stabilized by an RA53 (or similar) negative-temperature-coefficient (NTC) thermistor. TH_1 and RV_1 form a gain-determining feedback network. The thermistor is heated by the mean power output of the

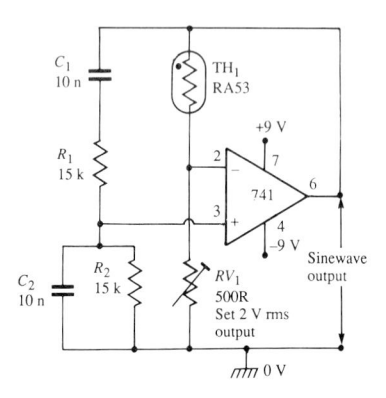

Figure 4.3 *Thermistor stabilized 1 kHz Wien bridge oscillator*

op-amp, and at the desired output signal level has a resistance value double that of RV_1, thus giving the op-amp a gain of × 3 and the overall circuit a gain of unity. If the oscillator output amplitude starts to rise, TH_1 heats up and reduces its resistance, or vice versa, thereby automatically reducing the gain of the circuit and stabilizing the amplitude of the output signal.

An alternative method of thermistor stabilization is shown in *Figure 4.4*. In this case a low-current lamp is used as a positive-temperature-coefficient thermistor, and is placed in the lower part of the gain-determining feedback network. Thus, if the output amplitude increases, the lamp heats up and increases its resistance, thereby reducing the circuit gain and providing automatic amplitude stabilization. This circuit also shows how the Wien network can be modified by using a twin-gang pot to make the oscillator frequency variable over the range 150 Hz to 1.5 kHz, and how the sine wave output amplitude can be made variable via RV_3.

In the *Figure 4.3* and *4.4* circuits, the pre-set pot should be adjusted to set

Waveform generator circuits 55

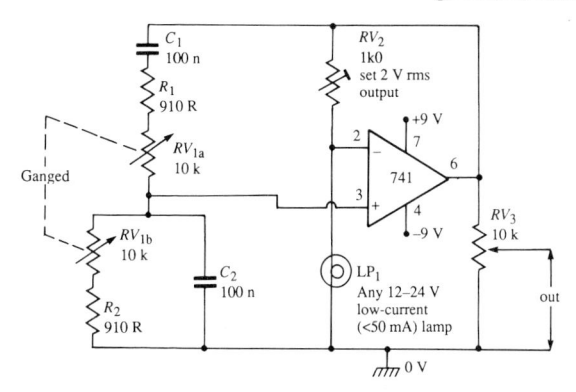

Figure 4.4 *150 Hz–1.5 kHz lamp-stabilized Wien bridge oscillator*

the maximum mean output level to about 2 V rms, and under this condition the sine wave has a total harmonic distortion of about 0.1%. Note that a slightly annoying feature of thermistor-stabilized circuits is that, in variable-frequency applications, the output amplitude of the sine wave tends to judder or bounce as the frequency control pot is swept up and down its range.

Diode stabilization

The amplitude bounce problem of variable-frequency circuits can be minimized by using the basic circuits of *Figures 4.5* or *4.6*, which rely on the onset of diode or zener conduction for automatic gain control. In essence, RV_2 is set so that the circuit gain is slightly greater than unity when the output is close to zero, causing the circuit to oscillate, but as each half-cycle nears the desired peak value one or other of the diodes starts to conduct and thus reduces the circuit gain, automatically stabilizing the peak amplitude of the output signal. This limiting technique typically results in the generation of 1% to 2% distortion on the sine wave output. The maximum peak-to-peak output of each circuit is roughly double the breakdown voltage of its diode regulator element.

In the *Figure 4.5* circuit, the diodes start to conduct at about 500 mV, so the circuit gives a peak-to-peak output of about 1V0. In the *Figure 4.6* circuit, zener diodes ZD_1 and ZD_2 are connected back-to-back, and may have values as high as 5V6, giving a peak-to-peak output of about 12 V. Each circuit is set up by adjusting RV_2 to the maximum value (minimum distortion) at which oscillation is maintained across the whole frequency band.

The frequency ranges of the above circuits can be changed via the C_1 and C_2 values; increasing them by a decade reduces the frequency values by a decade,

56 Waveform generator circuits

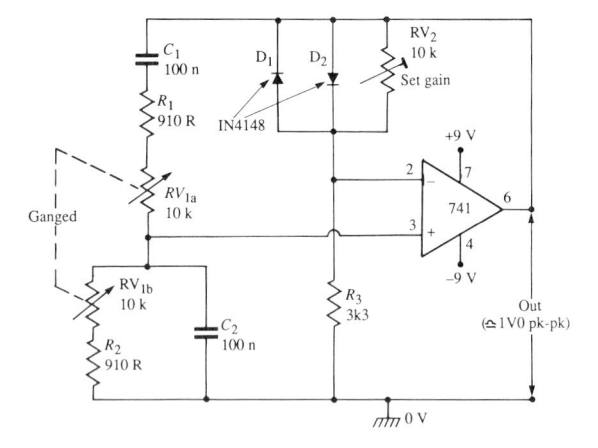

Figure 4.5 *Diode-regulated 150 Hz–1.5 kHz Wien bridge oscillator*

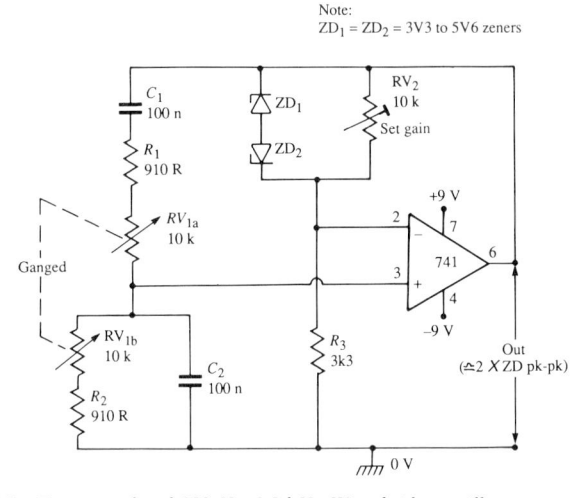

Figure 4.6 *Zener-regulated 150 Hz–1.5 kHz Wien bridge oscillator*

etc. *Figure 4.7* shows a variable-frequency Wien oscillator that covers the range 15 Hz to 15 kHz in three switched decade ranges. The circuit uses zener diode amplitude stabilization, and its output is adjustable via both switched and fully variable attenuators. Note that the maximum useful operating frequency of this type of circuit is restricted by the slew rate limitations of the op-amp; the limit is about 25 kHz with a 741 op-amp, or about 70 kHz with a CA3140.

Waveform generator circuits 57

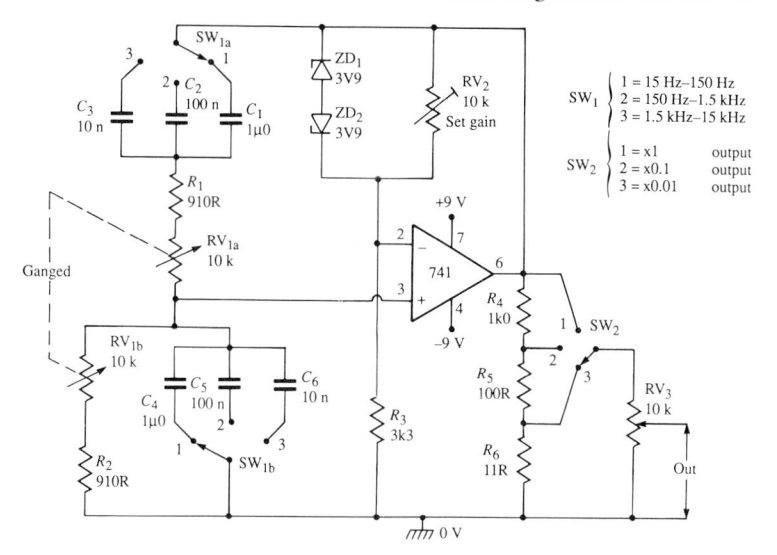

Figure 4.7 *Three-decade (15 Hz–15 kHz) Wien bridge oscillator*

Twin-T oscillators

Another way of making a sine wave oscillator is to wire a twin-T network between the output and input of an inverting op-amp, as shown in *Figure 4.8*. The twin-T network comprises R_1–R_2–C_3 and C_1–C_2–R_3–RV_1, and in a balanced circuit these components are in the ratios $R_1 = R_2 = 2(R_3 + RV_1)$, and $C_1 = C_2 = C_3/2$. When the network is perfectly balanced it acts as a frequency-dependent attenuator that gives zero output at a centre frequency, f_o, of $1/2\pi.R_1.C_1$, and a finite output at all other frequencies. When the network is imperfectly balanced it gives a minimal but finite output at f_o, and the phase of this output depends in the direction of the imbalance. If the imbalance is caused by $(R_3 + RV_1)$ being too low in value, the output phase is inverted relative to the input.

In *Figure 4.8*, the twin-T network is wired between the output and the inverting input of the op-amp, and RV_1 is critically adjusted so that the twin-T gives a small phase-inverted output at the f_o of 1 kHz; zero overall phase inversion thus occurs around the feedback loop, and the circuit oscillates at a centre frequency of 1 kHz. In practice, RV_1 is adjusted so that oscillation is barely sustained, and under this condition the sine wave output has less than 1% distortion. Automatic amplitude control occurs because of the progressive non-linearity of the op-amp as the output signal approaches clipping

58 Waveform generator circuits

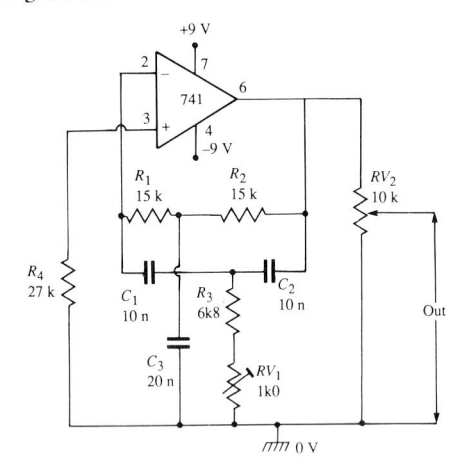

Figure 4.8 *1 kHz twin-T oscillator*

level. The output amplitude is fully variable from zero to about 5 V rms via RV_2.

Figure 4.9 shows a method of amplitude control that gives slightly less distortion. Here, D_1 provides a feedback signal via potential divider RV_2. This diode progressively conducts and reduces the circuit gain when the diode forward voltage exceeds 500 mV. To set up the circuit, first set RV_2 slider to the op-amp output and adjust RV_1 so that oscillation is just sustained; under

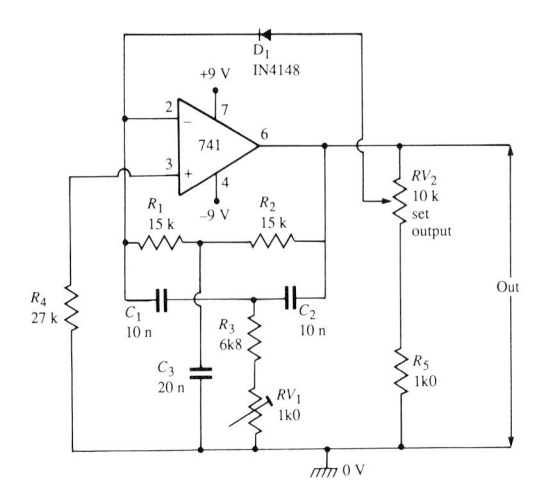

Figure 4.9 *Diode-regulated 1 kHz twin-T oscillator*

Waveform generator circuits 59

this condition the output signal has an amplitude of about 500 mV peak-to-peak. RV_2 then enables the output signal to be varied between 170 mV and 3 V rms.

Note that these twin-T circuits make good fixed-frequency oscillators, but are no good for variable-frequency use, due to the difficulties of simultaneously varying three or four network components.

Square wave generators

An op-amp can be used to generate square waves by using the basic relaxation oscillator configuration of *Figure 4.10*. The circuit uses dual power supplies, and its output switches alternately between the positive and negative saturation levels of the op-amp. Potential divider R_2–R_3 feeds a fraction of this voltage back to the non-inverting input of the op-amp, to provide the circuit with an 'aiming' voltage, and feedback components R_1–C_1 act as a time-constant network.

The basic operation of the *Figure 4.10* circuit is such that, when the output is high, C_1 charges up via R_1 until the C_1 voltage reaches the positive aiming value set by R_2–R_3, at which point a comparator action occurs and the op-amp output regeneratively switches negative, causing C_1 to start to discharge via R_1 until the C_1 voltage falls to the negative aiming value set by R_2–R_3, at which point the op-amp output regeneratively switches positive again, and the whole sequence then repeats ad infinitum. The action is such that a symmetrical square wave is developed at the output of the op-amp, and a none-linear triangle waveform is developed across C_1; these waveforms swing symmetrically either side of the zero-volts line. A fast op-amp, such as the CA3140, should be used if good output rise and fall times are needed from the square wave.

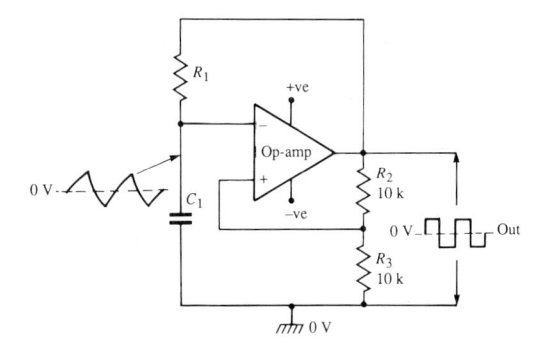

Figure 4.10 *Basic relaxation oscillator circuit*

60   Waveform generator circuits

Note that the operating frequency of this circuit can be varied by altering either the R_1 or C_1 values, or by altering the R_2–R_3 ratios; this circuit is thus quite versatile. *Figure 4.11* shows how it can be adapted to make a practical 500 Hz to 5 kHz square-wave generator, with frequency variation obtained by altering the attenuation ratio of R_2–RV_1–R_3. *Figure 4.12* shows how the circuit can be improved by using RV_2 to pre-set the range of the RV_1 frequency control, and by using RV_3 as an output amplitude control.

Figure 4.13 shows how the above circuit can be modified to make a general-purpose square wave generator that covers the 2 Hz to 20 kHz range in four switched decade ranges. Pre-set pots RV_1 to RV_4 are used to precisely set the minimum frequency of the 2 Hz–20 Hz, 20 Hz–200 Hz, 200 Hz–2 kHz, and 2 kHz–20 kHz ranges respectively.

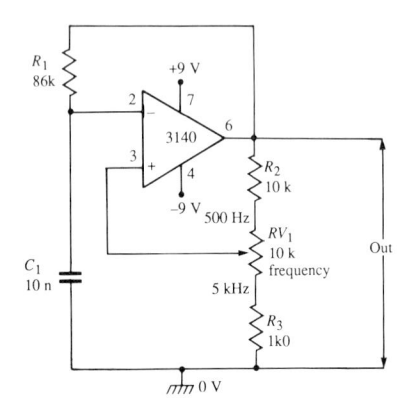

Figure 4.11   *Simple 500 Hz–5 kHz square wave generator*

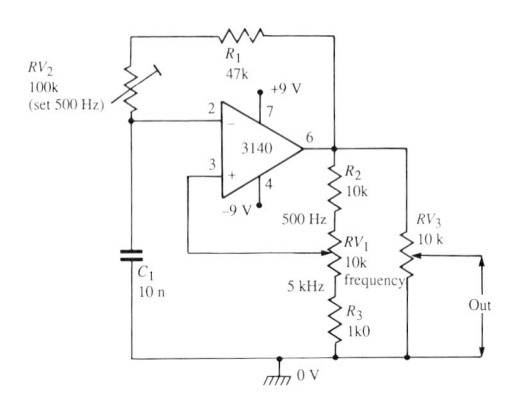

Figure 4.12   *Improved 500 Hz–5 kHz square wave generator*

Waveform generator circuits 61

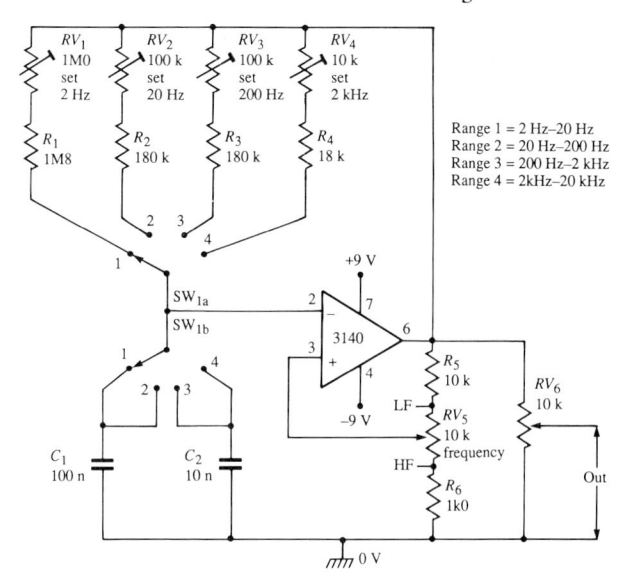

Figure 4.13 *4-decade, 2 Hz–20 kHz, square wave generator*

Variable symmetry

In the *Figure 4.10* circuit, C_1 alternately charges and discharges via R_1, and the circuit generates a symmetrical square wave output. The circuit can easily be modified to give a variable-symmetry output by providing C_1 with alternate charge and discharge paths, as shown in *Figures 4.14* and *4.15*.

In the *Figure 4.14* circuit the mark/space (M/S) ratio of the output waveform is fully variable from 11:1 to 1:11 via RV_1, and the frequency is variable from 650 Hz to 6.5 kHz via RV_2. The action is such that C_1 alternatively charges up via R_1–D_2 and the left-hand side of RV_1, and discharges via R_1–D_2 and the right-hand side of RV_1, giving a variable-symmetry output. Note that variation of RV_1 has negligible effect on the operating frequency of the circuit.

In the *Figure 4.15* circuit, the mark period is determined by C_1–D_1–R_1, and the space period by C_1–D_2–R_1; these periods differ by a factor of one hundred, so the circuit generates a narrow pulse waveform. The pulse frequency is variable from 300 Hz to 3 kHz via RV_1.

Resistance activation

Note from the description of the *Figure 4.10* oscillator that the circuit actually changes state in each half-cycle at the point where the C_1 voltage reaches the

62 Waveform generator circuits

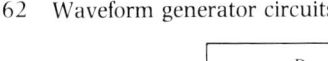

Figure 4.14 *Square-wave generator with variable M/S ratio and frequency*

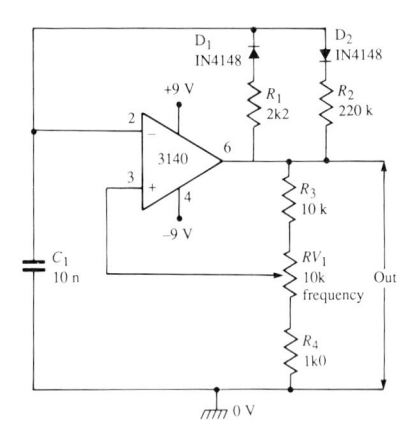

Figure 4.15 *Variable-frequency narrow-pulse generator*

'aiming' value set by the R_2–R_3 potential divider; if C_1 is unable to attain this voltage, the circuit will not oscillate. Thus, if the circuit is modified as shown in *Figure 4.16*, in which RV_1 is wired in parallel with C_1 and forms a potential divider with R_1, and R_2–R_3 have a 1:1 ratio, the circuit will oscillate only if RV_1 has a value greater than R_1. This circuit can thus function as a resistance-activated oscillator.

Figures 4.17 and *4.18* show two practical applications of the resistance-activated oscillator. The *Figure 4.17* circuit acts as a precision light-activated

Waveform generator circuits 63

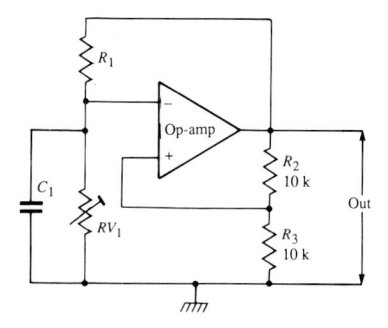

Figure 4.16 *Basic resistance-activated relaxation oscillator*

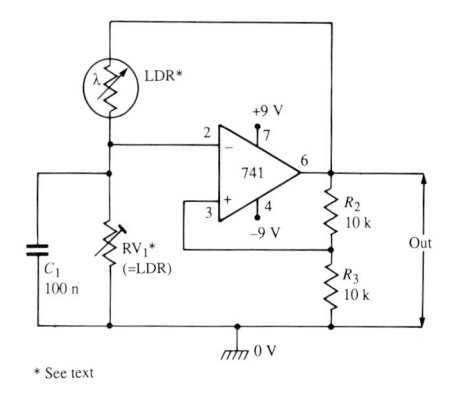

Figure 4.17 *Precision light-activated oscillator/alarm*

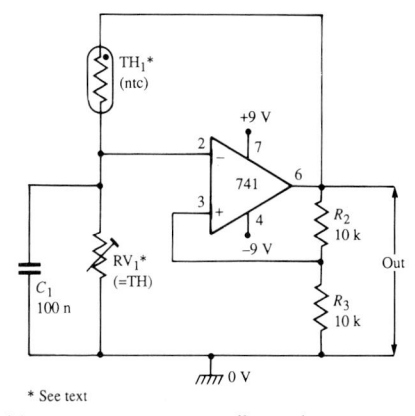

Figure 4.18 *Precision over-temperature oscillator/alarm*

64 Waveform generator circuits

oscillator or alarm, and uses an LDR as the resistance-activating element. The circuit can be converted to a dark-activated oscillator by transposing the LDR–RV_1 positions.

The *Figure 4.18* circuit is similar to the above, but uses NTC thermistor. TH_1 as the resistance-activating element, and acts as a precision over-temperature oscillator/alarm. The circuit can be converted to an under-temperature oscillator by transposing TH_1 and RV_1.

In the above two circuits the LDR or TH_1 can have any resistance in the range 2k0 to 2M0 at the required trigger level, and RV_1 must have the same value as the activating element at the desired trigger level. RV_1 sets the trigger level; the C_1 value can be altered to change the oscillator frequency.

Triangle/square generation

Figure 4.19 shows the basic circuit of a function generator that simultaneously generates a linear triangle and a square output waveform, and uses two op-amps. IC_1 is wired as an integrator, driven from the output of IC_2, and IC_2 is wired as a differential voltage comparator, driven from the output of IC_1 via potential divider R_2–R_3, which is connected between the outputs of IC_1 and IC_2. The square wave output of IC_2 switches alternately between positive and negative saturation. The circuit functions as follows.

Suppose initially that the output of IC_1 is positive and the output of IC_2 has just switched to positive saturation. The inverting input of IC_1 is a virtual earth point, so a current (*i*) of $+V_{sat}/R_1$ flows into R_1, causing the output of IC_1 to start to swing down linearly at a rate of i/C_1 volts per second. This output is fed, via the R_2–R_3 divider, to the non-inverting input of IC_2, which has its inverting terminal referenced directly to ground.

Consequently, the output of IC_1 swings linearly to a negative value until the R_2–R_3 junction voltage falls to zero, at which point IC_2 enters a regenerative switching phase, in which its output abruptly switches to negative saturation.

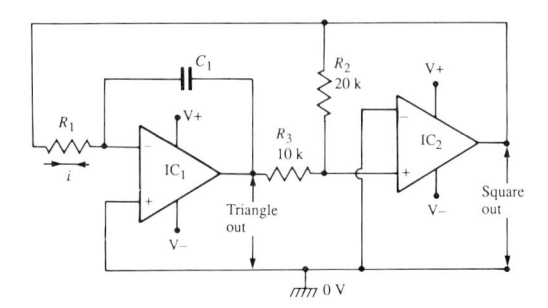

Figure 4.19 *Basic triangle-square function generator*

Waveform generator circuits 65

This reverses the inputs of IC_1 and IC_2, so IC_1 output starts to rise linearly, until it reaches a positive value at which the R_2–R_3 junction voltage reaches the zero volts reference value, initiating another switching action. The whole process then repeats add infinitum.

Important points to note about the *Figure 4.19* circuit are that the peak-to-peak amplitude of the linear triangle waveform is controlled by the R_2–R_3 ratio, and that the operating frequency of the circuit can be altered by changing either the ratios of R_2–R_3, the values of R_1 or C_1, or by feeding R_1 from a potential divider connected to the output of IC_2 (rather than directly from IC_2 output). *Figure 4.20* shows the practical circuit of a variable-frequency triangle/square generator using the latter technique.

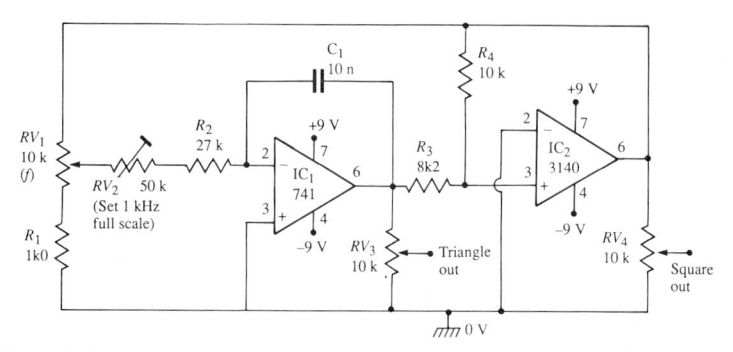

Figure 4.20 *100 Hz–1 kHz triangle-square function generator*

In *Figure 4.20*, the input current to C_1 (obtained from RV_2–R_2) can be varied over a 10:1 range via RV_1, enabling the frequency to be varied from 100 Hz to 1 kHz; RV_2 enables the full-scale frequency to be set to precisely 1 kHz. The amplitude of the linear triangle output waveform is fully variable via RV_3, and of the square wave via RV_4.

The *Figure 4.20* circuit generates symmetrical output waveforms, since C_1 alternately charges and discharges at equal current values (determined by RV_2–R_2, etc.). *Figure 4.21* shows how the above circuit can be modified to make a variable-symmetry ramp/rectangle generator, in which the slope of the ramp and the M/S ratio of the rectangle is variable via RV_2. C_1 alternately charges via R_2–D_1 and the upper half of RV_2, and discharges via R_2–D_2 and the lower half of RV_2.

Switching circuits

To conclude this chapter, *Figures 4.22* to *4.24* show three ways of using op-amps as simple regenerative switches. *Figure 4.22* shows the connections for

66 Waveform generator circuits

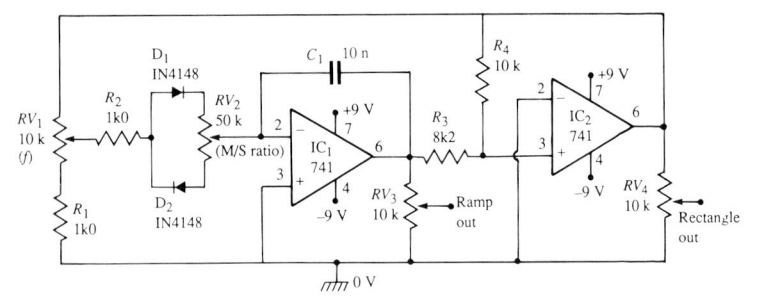

Figure 4.21 *100 Hz–1 kHz ramp-rectangle generator with variable slope – mark/space ratio*

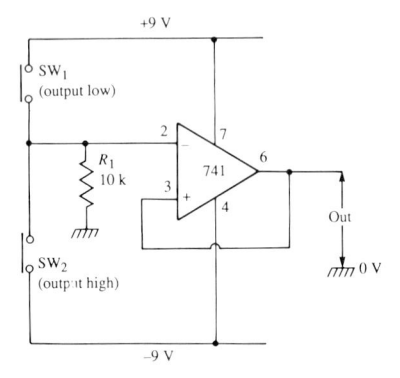

Figure 4.22 *Simple manually-triggered bistable*

making a simple manually triggered bistable circuit. Note here that the inverting terminal of the op-amp is tied to the ground via R_1, and the non-inverting input is tied directly to the output. The circuit operates as follows.

Normally, SW_1 and SW_2 are open. If SW_1 is briefly closed, the op-amp inverting terminal is momentarily pulled high and the output is driven to negative saturation. Consequently, when SW_1 is released again, the inverting terminal returns to zero volts but the output and the non-inverting terminal remain in negative saturation. The output remains in this state until SW_2 is briefly closed, at which point the op-amp output switches to positive saturation, and locks into this state until SW_1 is again operated. The circuit thus gives a bistable form of operation. *Figure 4.23* shows how the circuit can be modified for operation from a single-ended power supply; in this case the inverting terminal of the op-amp is biased to half-supply volts via R_1 and the R_2–R_3 potential divider.

Waveform generator circuits 67

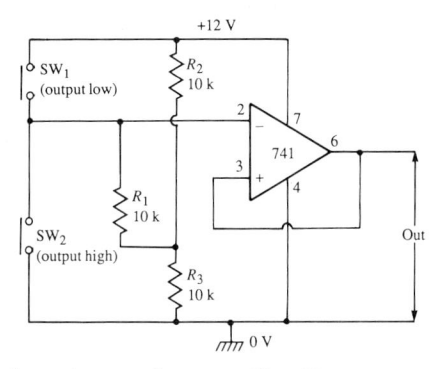

Figure 4.23 *Single-supply manually-triggered bistable*

Finally, *Figure 4.24* shows how to connect an op-amp as a Schmitt trigger, which can, for example, be used to convert a sine wave into a square wave output. The circuit operates as follows.

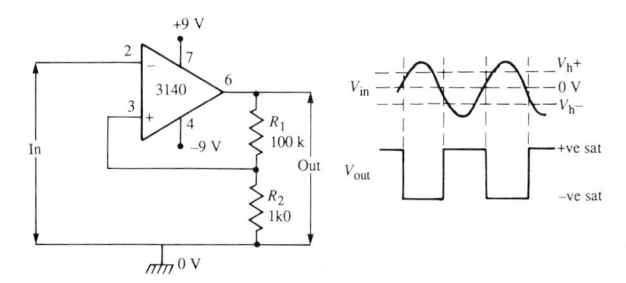

Figure 4.24 *Schmitt trigger*

Suppose initially that the op-amp output is at a positive saturation value of 8V0. Under this condition the R_1–R_2 divider feeds a positive reference voltage of 8V0 × $(R_1 + R_2)/R_2$ (about 80 mV in this case) to the non-inverting pin of the op-amp. Consequently, the output remains in this state until the input rises to a value equal to this voltage, at which point the op-amp output switches regeneratively to a negative saturation level of −8V0, feeding a reference voltage of −80 mV to the non-inverting input. The output then remains in this state until the input signal falls to −80 mV, at which point the op-amp output switches regeneratively back to the positive saturation level. The switching levels can be altered by changing the R_1 value.

5 Instrumentation and test-gear circuits

Op-amps can be used in a variety of instrumentation and test-gear applications. They can easily be used as precision rectifiers, peak voltage detectors, as ac/dc converters, and as fixed or variable voltage/power regulators. They can be used to convert standard dc digital voltmeter (DVM) modules into multi-range instruments capable of reading ac voltage or current, or resistance. When used in conjunction with moving coil meters they can be used to make dc and ac voltmeters, micro-ammeters, and linear-scale ohmmeters, etc. Practical versions of all of these circuits are described in this chapter.

Electronic rectifiers

Conventional diodes are poor rectifiers of low-level ac signals, because they do not start to conduct until the applied voltage exceeds a certain knee value; silicon diodes have knee values of about 600 mV, and thus give negligible rectification of signal voltages below this value. This weakness can be effectively overcome by wiring the diode into the feedback loop of an op-amp, in such a way that the effective knee voltage is reduced by a factor equal to the op-amp's open-loop voltage gain; the combination then acts as a near-perfect rectifier that can respond to signal inputs as low as a fraction of a millivolt. *Figure 5.1* shows a simple half-wave rectifier of this type.

The *Figure 5.1* circuit is actually wired as a unity-gain voltage follower, with silicon diode D_1 connected into its negative feedback loop, and with the circuit output taken from across load resistor R_1. When positive input signals are applied to the circuit the op-amp output also goes positive, thus forward

Instrumentation and test gear circuits   69

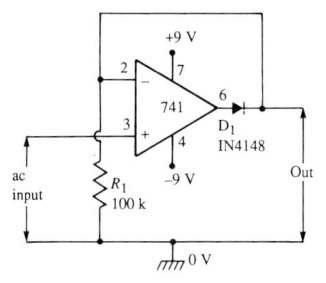

Figure 5.1 *Simple half-wave rectifier circuit*

biasing D_1 and feeding a signal current into R_1, which thus accurately follows the input signal. When negative signals are applied to the circuit the op-amp output also goes negative, but under this condition D_1 is reverse biased and is thus unable to feed current into R_1, which thus gives zero output. The circuit thus follows positive input signals but rejects negative ones, and thus has the characteristics of a near-perfect rectifier.

Figure 5.2 shows how the above circuit can be modified to act as a peak voltage detector by wiring C_1 in parallel with R_1. This capacitor charges rapidly, via D_1, to the peak positive value of an input signal, but discharges slowly via R_1 when the signal falls below the peak value. IC_2 is used as a voltage-following buffer stage, to ensure that R_1 is not shunted by external loading effects.

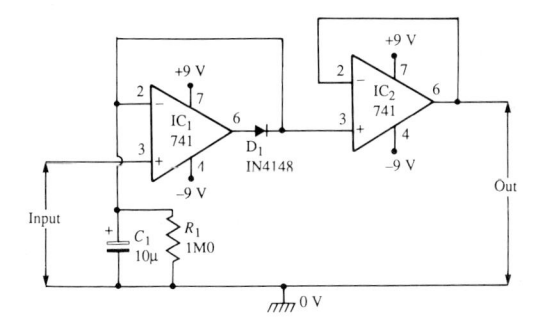

Figure 5.2 *Peak detector with buffered output*

Note that the *Figure 5.1* and *5.2* circuits each have a very high input impedance.

Precision rectifiers

The *Figure 5.1* rectifier circuit has a rather limited frequency response, and may produce a slight negative output signal if D_1 has poor reverse resistance characteristics. *Figure 5.3* shows an alternative type of half-wave rectifier circuit, which has a greatly improved rectifier performance at the expense of a greatly reduced input impedance.

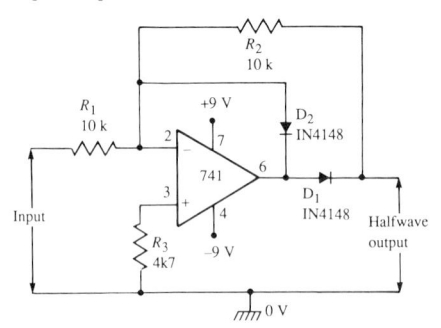

Figure 5.3 *Precision half-wave rectifier*

In *Figure 5.3* the op-amp is wired as an inverting amplifier with a 10 k (R_1) input impedance. When the input signal is negative the op-amp output swings positive, forward biasing D_1 and developing an output across R_2. Under this condition the voltage gain equals $(R_2 + R_D)/R_1$, where R_D is the active resistance of this diode. Thus, when D_1 is operating below its knee value its resistance is large and the circuit gives high gain, but when D_1 is operating above the knee value its resistance is very low and the circuit gain equals R_2/R_1. The circuit thus acts as an inverting precision rectifier to negative input signals.

When the input signal goes positive the op-amp output swings negative, but the negative swing is limited to -600 mV via D_2, and the output at the D_1-R_2 junction does not significantly shift from zero under this condition. This circuit thus produces a positive-going half-wave rectified output. Note that the circuit can be made to give a negative-going half-wave rectified output by simply reversing the polarities of the two diodes.

Figure 5.4 shows how a negative-output version of the above circuit can be combined with an inverting adder to make a precision full-wave rectifier. Here, IC_2 inverts and gives $\times 2$ gain (via R_3-R_5) to the half-wave rectified signal of IC_1, and inverts and gives unity gain (via R_4-R_5) to the original input signal (E_{in}). Thus, when negative input signals are applied, the output of IC_1 is zero, so the output of IC_2 equals $+E_{in}$. When positive input signals are applied, IC_1 gives a negative output, so IC_2 generates an output of $+2E_{in}$ in via IC_1 and $-E_{in}$ via the original input signal, thus giving an actual output of

Instrumentation and test gear circuits 71

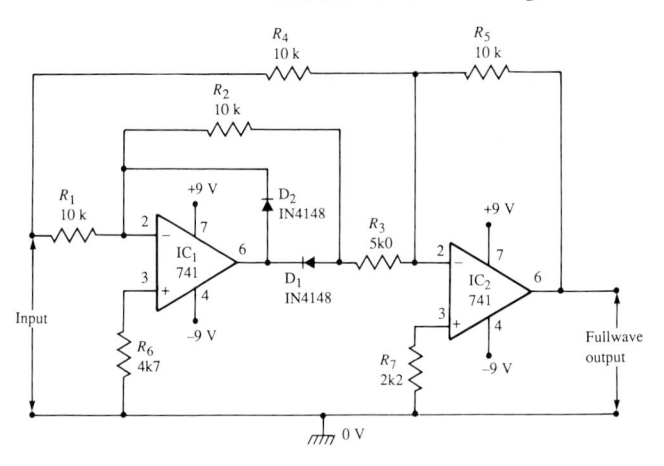

Figure 5.4 *Precision full-wave rectifier*

$+E_{in}$. The output of this circuit is thus positive, and always has a value equal to the absolute value of the input signal.

ac/dc converters

The *Figure 5.3* and *5.4* circuits can be made to function as precision ac/dc converters by first providing them with voltage-gain values suitable for form-factor correction, and by then integrating their outputs to give the ac/dc conversion, as shown in *Figures 5.5* and *5.6* respectively. Note that these circuits are intended for use with sine wave input signals only.

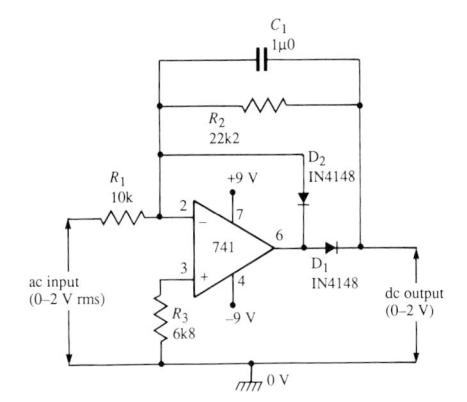

Figure 5.5 *Precision half-wave ac/dc converter*

72 Instrumentation and test gear circuits

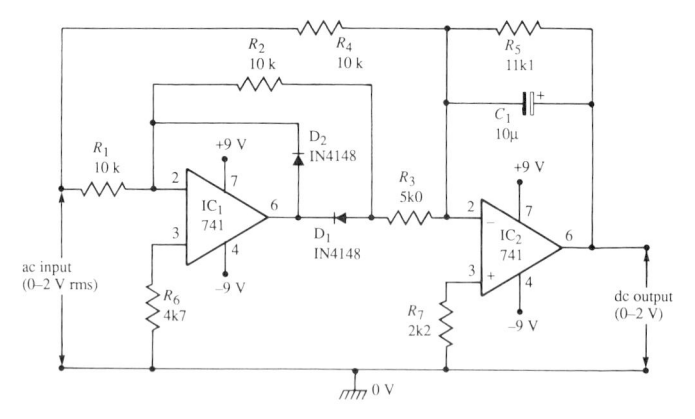

Figure 5.6 *Precision full-wave ac/dc converter*

In the half-wave ac/dc converter of *Figure 5.5* the circuit gives a voltage gain of × 2.22 via R_2/R_1, to give form-factor correction, and integration is accomplished via C_1–R_2. Note that this circuit has a high output impedance, and the output must be buffered if it is to be fed to low-impedance loads.

In the full-wave ac/dc converter of *Figure 5.6*, the circuit has a voltage gain of × 1.11 to give form-factor correction, and integration is accomplished via C_1–R_5. This circuit has a low-impedance output.

DVM converter circuits

Precision $3\frac{1}{2}$-digit digital voltmeter (DVM) modules are readily available at modest cost, and can easily be used as the basis of individually built multi-range and multi-function meters. These modules are usually powered via a 9 V battery, and have a basic full-scale measurement sensitivity of 200 mV dc and a near-infinite input resistance. They can be made to act as multi-range dc voltmeters by simply feeding the test voltage to the module via a suitable 'multiplier' (resistive attenuator) network, or as multi-range dc current meters by feeding the test current to the module via a switched current shunt.

A DVM module can be used to measure ac voltages by connecting a suitable ac/dc converter to its input terminals, as shown in *Figure 5.7*. This particular converter has a near-infinite input impedance. The op-amp is used in the non-inverting mode, with dc feedback applied via R_2 and ac feedback applied via C_1–C_2 and the diode-resistor network. The converter gain is variable over a limited range (to give form-factor correction) via RV_1, and the circuit's rectified output is integrated via R_6–C_3, to give dc conversion. The common terminal of the DVM module is internally biased at about 2.8 V

Instrumentation and test gear circuits 73

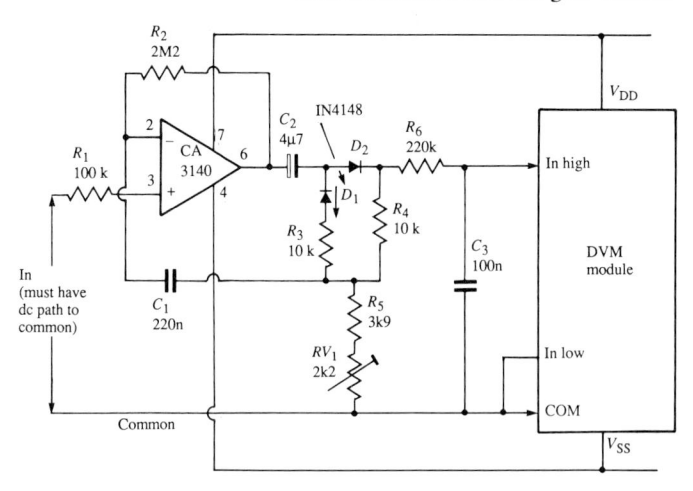

Figure 5.7 *ac/dc converter for use with DVM module*

below the V_{DD} (positive supply terminal) voltage, and the CA3140 op-amp uses the V_{DD}, common and V_{SS} teminals of the module as its supply rail points.

Figure 5.8 shows a simple frequency-compensated attenuator network used in conjunction with the above ac/dc converter to convert a standard DVM

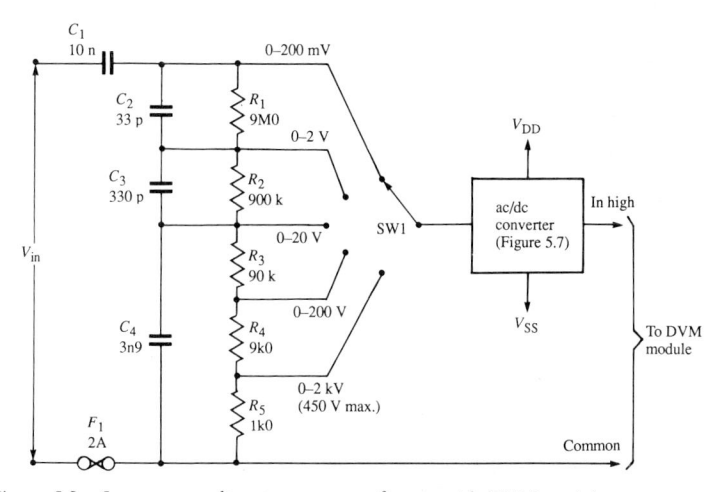

Figure 5.8 *5-range ac volt-meter converter for use with DVM modules*

74 Instrumentation and test gear circuits

module into a 5-range ac voltmeter, and *Figure 5.9* shows how a switched shunt network can be used to convert the module into a 5-range ac current meter.

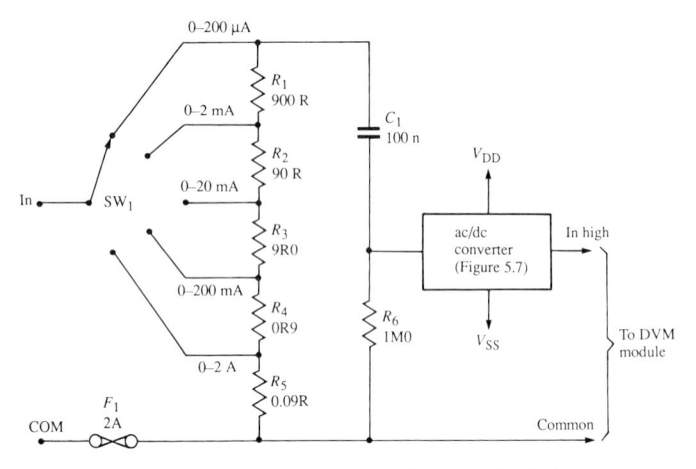

Figure 5.9 5-range ac current-meter converter for use with DVM modules

Figure 5.10 shows a circuit that can be used to convert a DVM module into a 5-range ohmmeter. This circuit actually functions as a multi-range constant-current generator, in which the constant current feeds (from Q_1 collector) into R_x, and the resulting R_x volt drop (which is directly proportional to the R_x value) is read by the DVM module.

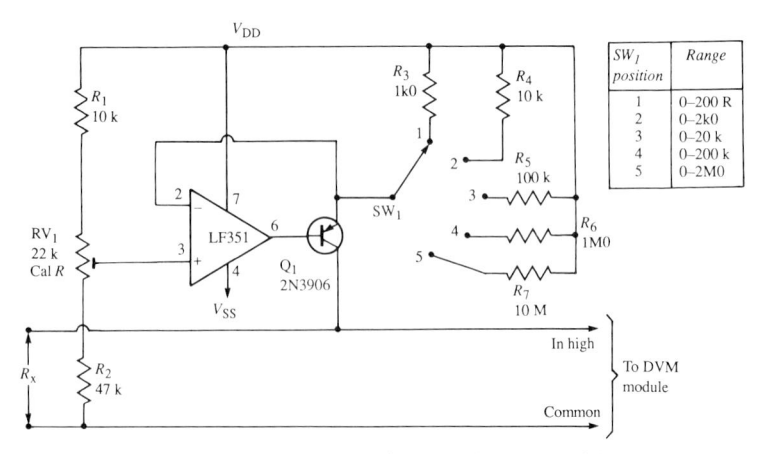

SW_1 position	Range
1	0–200 R
2	0–2k0
3	0–20 k
4	0–200 k
5	0–2M0

Figure 5.10 5-range ohmmeter converter for use with DVM modules

Instrumentation and test gear circuits 75

Here, Q_1 and the op-amp are wired as a compound voltage follower, in which the Q_1 emitter precisely follows the voltage set on RV_1 slider. In practice, this voltage is set at exactly 1 V below V_{DD}, and the emitter and collector (R_x) currents of Q_1 thus equal 1V0 divided by the R_3 to R_7 range-resistor value, e.g., 1 mA with R_3 in circuit, etc. The actual DVM module reads full scale when the R_x voltage equals 200 mV, and this reading is obtained when R_x has a value one-fifth of that of the range resistor, e.g., 200 R on range 1, or 2M0 on range 5, etc.

Analogue meter circuits

An op-amp can easily be used to convert a standard moving coil meter into a sensitive analogue voltage, current, or resistance meter, as shown in the practical circuits of *Figures 5.11 to 5.16*. All six circuits operate from dual 9-V supplies and are designed around the LF351 JFET op-amp, which has a very high input impedance and good drift characteristics. All circuits have an offset nulling facility, to enable the meter readings to be set to precisely zero with zero input, and are designed to operate with a moving coil meter with a basic sensitivity of 1 mA fsd.

If desired, these circuits can be used in conjunction with the 1 mA dc range of an existing multi-meter, in which case these circuits function as range converters. Note that each circuit has a 2k7 resistor wired in series with the output of its op-amp, to limit the available output current to a couple of milliamps and thus provide the meter with automatic overload protection.

Figure 5.11 shows a simple way of converting the 1 mA meter into a fixed-range dc millivolt meter with a full-scale sensitivity of 1 mV, 10 mV, 100 mV or 1 V. The circuit has an input sensitivity of 1 MΩ/V, and the table shows the appropriate R_1 value for different fsd sensitivities. To set the circuit up

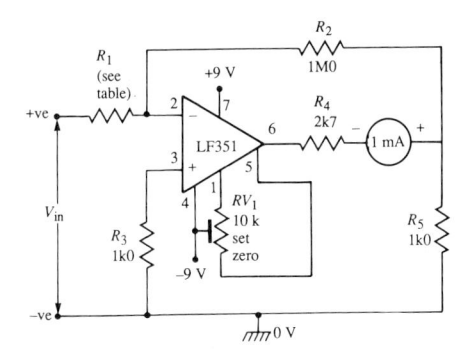

Figure 5.11 *A dc millivoltmeter circuit*

76 Instrumentation and test gear circuits

initially, short its input terminals together and adjust RV_1 to give zero deflection on the meter. The circuit is then ready for use.

Figure 5.12 shows a circuit that can be used to convert a 1 mA meter into either a fixed-range dc voltmeter with any full-scale sensitivity in the range 100 mV to 1000 V, or a fixed-range dc current meter with a full-scale sensitivity in the range 1 µA to 1 A. The table shows alternative R_1 and R_2 values for different ranges.

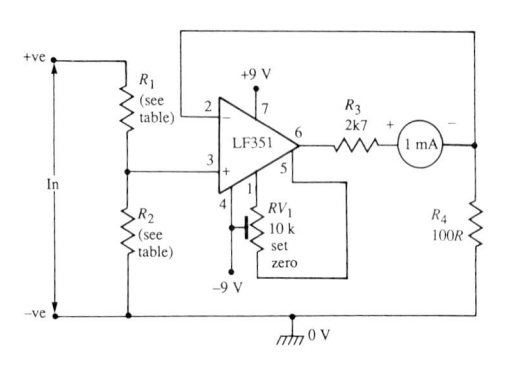

Voltmeter		
f.s.d.	R1	R2
1000 V	10M	1k0
100 V	10M	10 k
10 V	10M	100 k
1 V	900 k	100 k
100 mV	—	100 k

Current meter		
1 A	—	0R1
100 mA	—	1R0
10 mA	—	10R
1 mA	—	100R
100 µA	—	1k0
10 µA	—	10 k
1 µA	—	100 k

Figure 5.12 *A dc volt or current meter*

Figure 5.13 shows how the above circuit can be modified to make a 4-range dc millivolt meter with fsd ranges of 1 mV, 10 mV, 100 mV and 1V0, and *Figure 5.14* shows how it can be modified to make a 4-range dc micro-

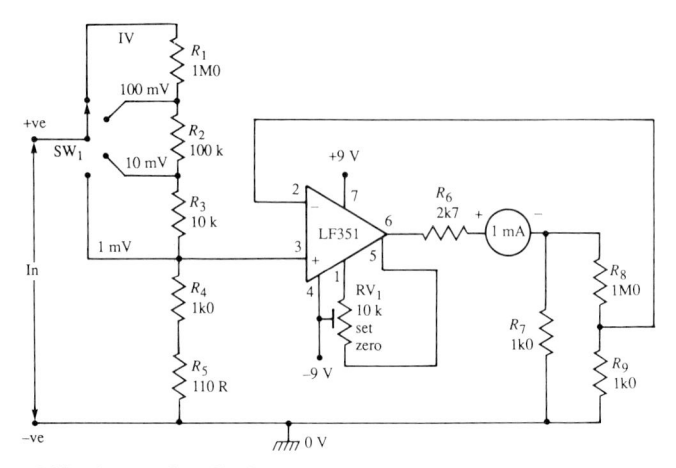

Figure 5.13 *4-range dc millivoltmeter*

Instrumentation and test gear circuits 77

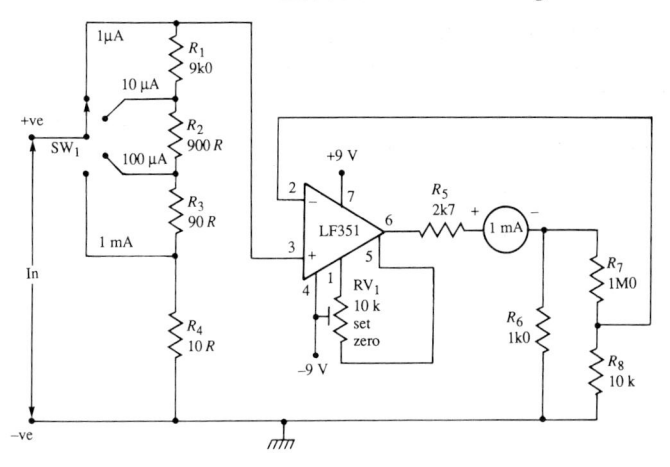

Figure 5.14 4-range dc microammeter

ammeter with fsd ranges of 1 µA, 10 µA, 100 µA and 1 mA. The range resistors used in these circuits should have accuracies of 2% or better.
Figure 5.15 shows the circuit of a simple but very useful 4-range ac millivoltmeter. The input impedance of the circuit is equal to R_1, and varies from 1k0 in the 1 mV fsd mode to 1M0 in the 1 V fsd mode. The circuit gives a useful performance at frequencies up to about 100 kHz when used in the 1 mV to 100 mV fsd modes. In the 1 V fsd mode the frequency response extends up to a few tens of kHz. This good frequency response is ensured by the LF351 op-amp, which has very good bandwidth characteristics.

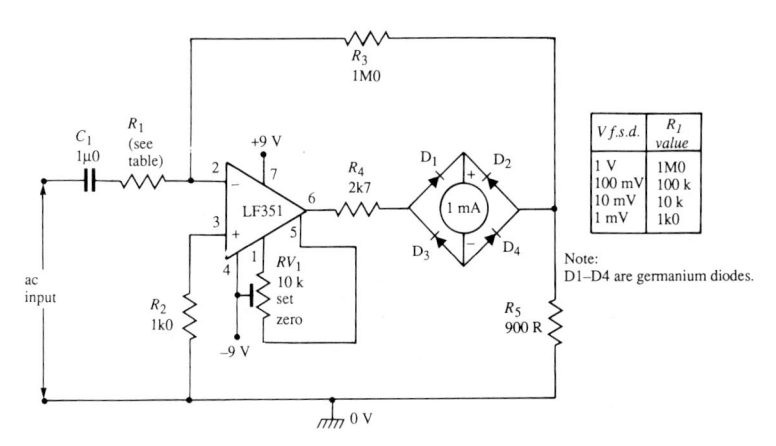

Figure 5.15 4-range ac millivoltmeter

78 Instrumentation and test gear circuits

Finally, *Figure 5.16* shows the circuit of a 5-range linear-scale ohmmeter, which has full-scale sensitivities ranging from 1k0 to 10 M. Range resistors R_5 to R_9 determine the measurement accuracy. Q_1–ZD_1 and the associated components simply apply a fixed 1 V (nominal) to the common side of the range-resistor network, and the gain of the op-amp circuit is determined by the ratios of the selected range-resistor and R_x and equals unity when these components have equal values: the meter reads full-scale under this condition, since it is calibrated to indicate full-scale when 1 V (nominal) appears across the R_x terminals.

To initially set up the *Figure 5.16* circuit, set SW_1 to the 10 k position and short the R_x terminals together. Then adjust the RV_1 set zero control to give zero deflection on the meter. Next, remove the short, connect an accurate 10 k resistor in the R_x position, and adjust RV_2 to give precisely full-scale deflection on the meter. The circuit is then ready for use, and should need no further adjustment for several months.

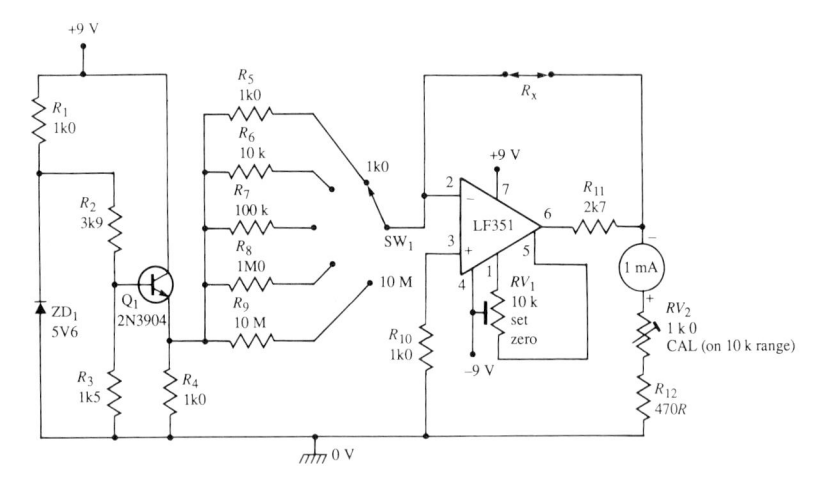

Figure 5.16 *5-range linear-scale ohmmeter*

Voltage reference circuits

An op-amp can be used as a fixed or variable voltage reference by wiring it as a voltage follower and applying a suitable reference to its input. An op-amp has a very high input impedance when used in the follower mode and thus draws near-zero current from the input reference, but has a very low output impedance and can supply several milliamps of current to an external load. Variations in output loading cause little change in the output voltage value.

Instrumentation and test gear circuits 79

Figure 5.17 shows a practical positive voltage reference with an output fully variable from +0.2 V to +12 V via RV_1. Zener diode ZD_1 generates a stable 12 V, which is applied to the non-inverting input of the op-amp via RV_1. A CA3140 op-amp is used here because its input and output can track signals to within 200 mV of the negative supply rail voltage. The complete circuit is powered from an unregulated single-ended 18 V supply.

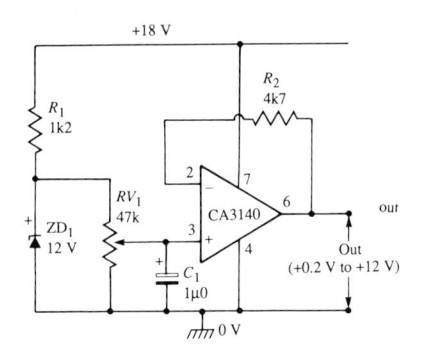

Figure 5.17   *Variable positive voltage reference*

Figure 5.18 shows a negative voltage reference that gives an output fully variable from −0.5 V to −12 V via RV_1. An LF351 op-amp is used in this design, because its input and output can track signals to within about 0.5 V of the positive supply rail value. Note that the op-amps used in these two regulator circuits are wide-band devices, and R_2 is used to enhance their circuit stability.

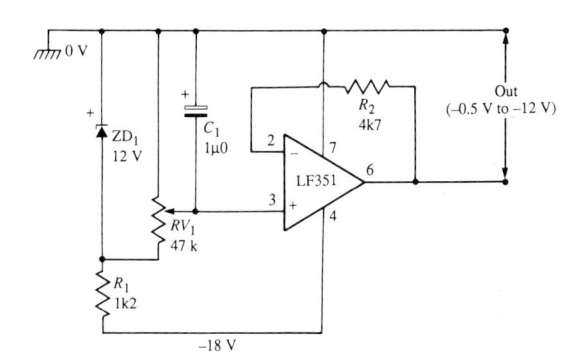

Figure 5.18   *Variable negative voltage reference*

Voltage regulator circuits

The basic circuits of *Figures 5.17* and *5.18* can be made to act as high-current regulated voltage (power) supply circuits by wiring current-boosting transistor networks into their outputs. *Figure 5.19* shows how the *Figure 5.17* circuit can be modified to act as a 1 V to 12 V variable power supply with an output current capability (limited by Q_1's power rating) of about 100 mA. Note that the base-emitter junction of Q_1 is included in the circuit's negative feedback loop, to minimize offset effects. The circuit can be made to give an output that is variable all the way down to zero volts by connecting pin-4 of the op-amp to a supply that is at least 2 V negative.

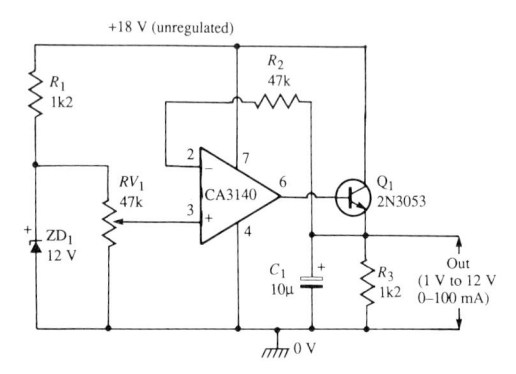

Figure 5.19 *Simple variable-voltage regulated power supply*

Figure 5.20 shows an alternative type of power supply circuit, in which the output is variable from 3 V to 15 V at currents up to 100 mA. In this case a fixed 3 V reference is applied to the non-inverting input terminal of the 741 op-amp via ZD_1 and the R_2–C_1–R_3 network, and the op-amp plus Q_1 are wired as a non-inverting amplifier with gain variable via RV_1. When RV_1 slider is set to the upper position, the circuit gives unity gain and gives an output of 3 V; when RV_1 slider is set to the lower position the circuit gives a gain of × 5 and thus gives an output of 15 V. The gain is fully variable between these two values. RV_2 enables the maximum output voltage to be pre-set to precisely 15 V.

Figure 5.21 shows how the above circuit can be modified to acts as a 3 V to 30 V, 0 to 1 A stabilized power supply unit (PSU). Here, the available output current is boosted by the Darlington-connected Q_1–Q_2 pair of transistors, the circuit gain is fully variable from unity to × 10 via RV_1, and the stability of the 3 V reference input to the op-amp is enhanced by the ZD_1 pre-regulator network.

Instrumentation and test gear circuits   81

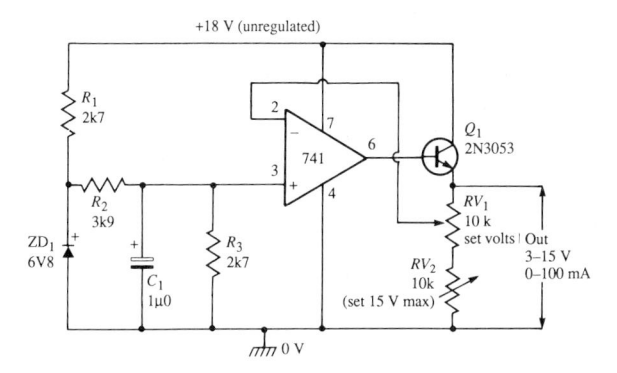

Figure 5.20   *3 V to 15 V, 0 to 100 mA stabilized PSU*

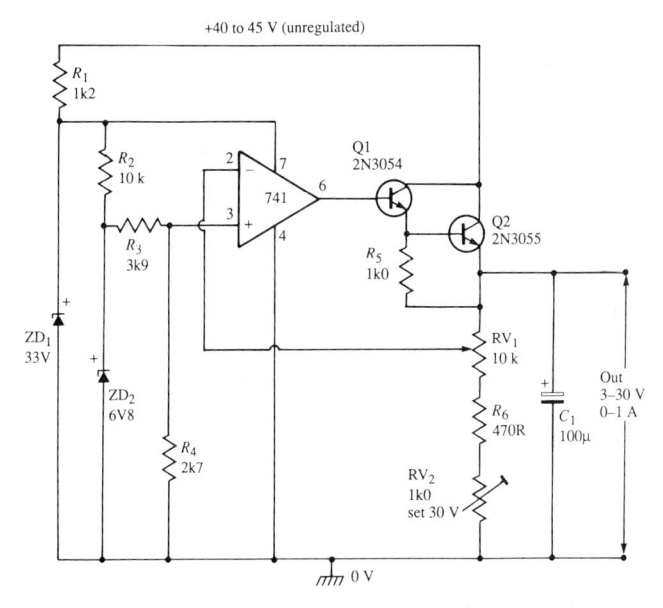

Figure 5.21   *3 V to 30 V, 0 to 1 A stabilized PSU*

Figure 5.22 shows how the above circuit can be further modified to incorporate automatic overload protection. Here, R_6 senses the magnitude of the output current and when this exceeds 1 A the resulting volt drop starts to bias Q_3 on, thereby shunting the base-drive current of Q_1 and automatically limiting the circuits output current.

82 Instrumentation and test gear circuits

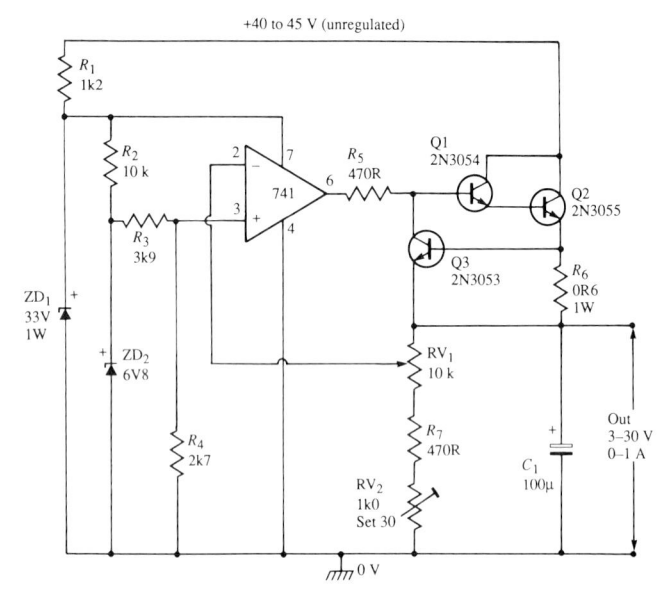

Figure 5.22 *3 V to 30 V stabilized PSU with overload protection*

Finally, to complete this chapter, *Figure 5.23* shows the circuit of a simple centre-tapped 0 to 30 V PSU that can provide maximum output currents of about 50 mA. The PSU has three output terminals, and can provide either 0 to +15 V between the common and positive terminals and 0 to −15 V between the common and negative terminals, or 0 to 30 V between the negative and positive terminals. The circuit operates as follows.

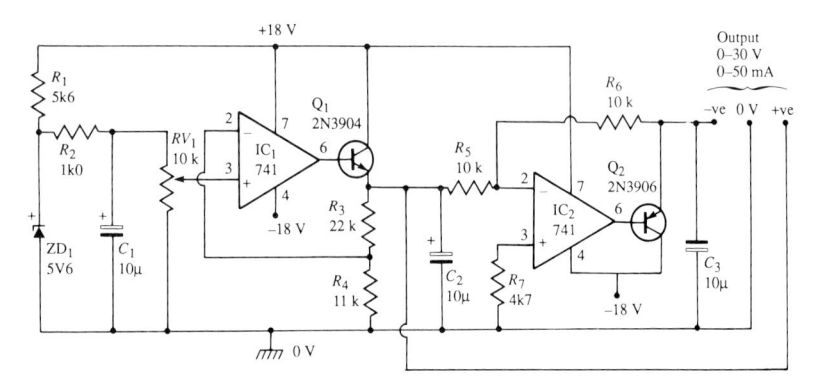

Figure 5.23 *Simple centre-tapped 0 to 30 V PSU*

ZD_1 and R_2–RV_1 provide a regulated 0 to 5 V potential to the input of IC_1. IC_1 and Q_1 are wired as a ×3 non-inverting amplifier, and thus generate a fully variable 0 to 15 V on the positive terminal of the PSU. This voltage is also applied to the input of the IC_2–Q_2 circuit, which is wired as a unity-gain inverting amplifier and thus generates an output voltage of identical magnitude but opposite polarity on the negative terminal of the PSU. The output current capability of each terminal is limited to about 50 mA by the power ratings of Q_1 and Q_2, but can easily be increased by replacing these components with Darlington (super-alpha) power transistors of appropriate polarity.

6 Compound circuits

The performance of a standard op-amp can be greatly improved or modified by connecting one or more transistors into its output feedback network, to make a compound or hybrid op-amp. Such circuits offer a low cost way of getting characteristics such as very high slew rates, or high output currents, or output voltage swings of up to hundreds of volts, which are not economically available from conventional op-amps.

The output current of an op-amp such as the CA3140, for example, is limited to only a few milliamperes, but can easily be boosted to several amps by compounding it with a couple of power transistors. Again, the output of the CA3140 normally swings to within only 2 V of the positive supply rail voltage, is limited to a maximum swing of about 32 V, and has a slew rate limit of about 9 V, but if the op-amp is coupled to a single common-emitter transistor the output of the resulting compound op-amp can easily swing within a few tens of millivolts of the supply rails, can have a swing of up to several hundred volts, and can have a slew rate of 100 $V/\mu s$.

Compound op-amp design is rarely mentioned in the technical press, but is of great practical value. The subject is fairly large, with hundreds of op-amp/transistor combinations being possible, so for the sake of simplicity only compound designs intended for use in single-ended supply applications and based on the CA3140 op-amp (which has input and output terminals that can swing all the way down to the negative supply rail) are described in this chapter. The principles described can, however, easily be adapted for use with other types of op-amp and with alternative supply arrangements.

Current boosting

The best known application of the compound op-amp is as a boosted-current voltage follower. *Figure 6.1* shows the CA3140 wired as a normal (unboosted)

Compound circuits 85

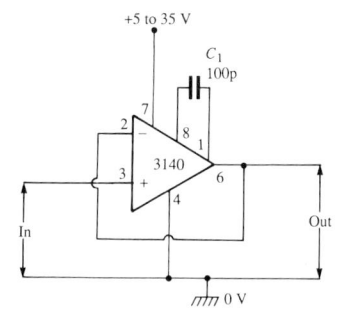

Figure 6.1 *This standard op-amp voltage follower circuit can supply only a few milliamperes of output current*

voltage follower or unity-gain non-inverting amplifier, in which the output is shorted to the inverting input terminal, giving 100% negative feedback. This circuit gives a near-infinite input impedance and a very low output impedance, and its output 'follows' the input signal to within a few millivolts of the zero-voltage rail and to within a couple of volts of the positive supply rail. C_1 is wired between pins 1 and 8 to reduce the op-amp's slew rate and thus enhance circuit stability.

A weakness of this circuit is that its output currents are limited to a few milliamperes by the op-amp's internal circuitry. This weakness can be overcome by compounding the op-amp with a transistor emitter follower stage, as shown in *Figure 6.2*. Note that the base-emitter junction of Q_1 is wired into the op-amp's negative feedback loop, thus eliminating the effects of junction non-linearity and offset voltage, and that zero phase shift occurs

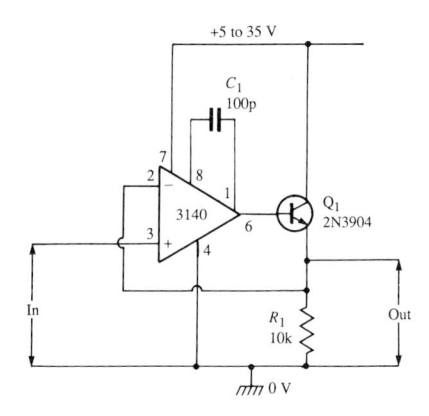

Figure 6.2 *Compound op-amp with boosted output current*

between Q_1 base and emitter. The output of the circuit (taken from Q_1 emitter) thus still precisely follow the input signal, but its output current is limited only by the ratings of Q_1.

In practice, the safe output current of this circuit is limited to about 50 mA by the power rating of Q_1. The available current can be boosted to an amp or so by replacing Q_1 with a Darlington pair of transistors, as shown in *Figure 6.3*, Q_2 being a high power device. The base-emitter junctions of both transistors are wired into the op-amp's feedback loop.

The *Figure 6.2* and *6.3* circuits give unidirectional current boosting, i.e., they can source (give out) high currents (via the output transistors), but can sink (take in) only relatively low ones (via the emitter-to-ground resistor of the output transistor); they are thus not suitable for use with AC-coupled low-impedance output loads. This weakness can be overcome by using a complementary emitter follower as the output stage, thus enabling the circuit to give bidirectional output currents. A very simple circuit of this type was described in Chapter 2 (*Figure 2.13*).

Figure 6.4 shows an improved version of the bidirectional circuit, in which D_1–D_2 provide a degree of offset biasing to the two output transistors, to minimize crossover distortion problems. This circuit can supply maximum mean output currents of about 50 mA. It can source currents via Q_1, and can sink them via Q_2.

Compound voltage followers

The current-boosting techniques shown in *Figures 6.2* to *6.4* are probably well known to most readers. They rely on the fitting of an additional non-inverting

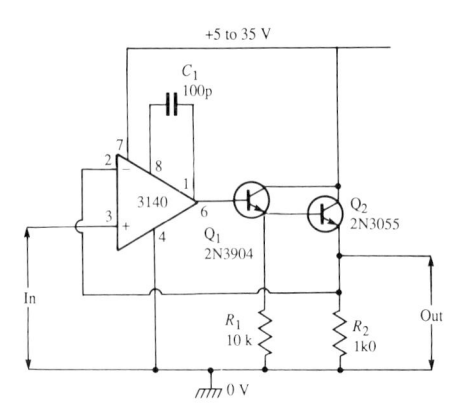

Figure 6.3 *This version of the follower can supply output currents of up to 1 A*

Compound circuits 87

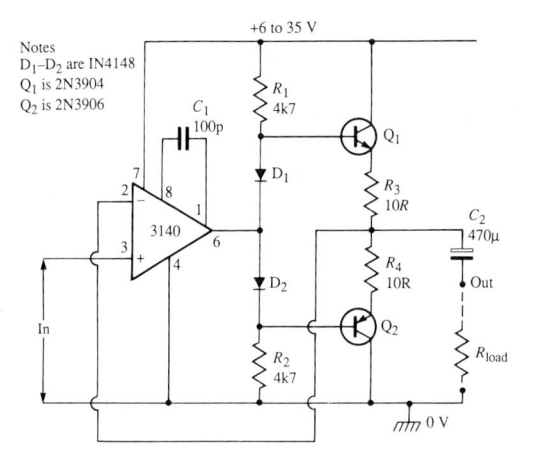

Notes
D_1–D_2 are IN4148
Q_1 is 2N3904
Q_2 is 2N3906

Figure 6.4 *Compound op-amp with boosted bidirectional output current*

stage to an existing op-amp, and are very easy to understand. In the remainder of this chapter, however, we describe unusual compounding techniques that rely on the fitting of additional *inverting* stages to op-amps, and these techniques are slightly less easy to understand. We start off by showing how these techniques can be used to make high-performance voltage follower circuits.

Figure 6.5 shows the circuit of a compound voltage follower or unity-gain non-inverting amplifier, in which common emitter transistor Q_1 is wired to the op-amp output and has its collector wired into the negative feedback loop of the circuit. This design should be compared with the 'standard' circuit of *Figure 6.1*. Note in *Figure 6.5* that, since an additional inverting stage (Q_1) has

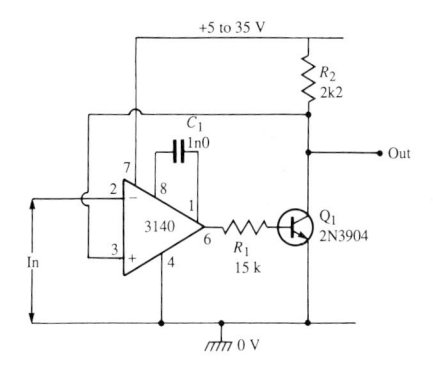

Figure 6.5 *Basic compound 'follower' circuit*

88 Compound circuits

been added to the op-amp, the input terminal notations of the 3140 must be mentally transposed to understand circuit operation.

Thus, the input signal is applied to the 3140 terminal marked 'inverting' (pin 2), which now acts as the non-inverting terminal of the compound op-amp, and the feedback connection from the compound output (Q_1 collector) goes to the 3140 terminal marked as non-inverting (pin 3), which now acts as the inverting terminal of the compound op-amp. The *Figure 6.5* circuit is thus, in theory, identical to that of *Figure 6.1*, and in fact gives a virtually identical performance, as follows:

When the input of the *Figure 6.5* circuit is at zero volts the 3140 drives Q_1 on and pulls its collector to saturation (typically within 50 mV of zero volts). When the input is above zero but at least 2 V below the supply rail value, the 100% negative feedback of the circuit forces Q_1 collector to take up a value identical to the 3140 input signal. The basic 3140 cannot follow signals that are within 2 V of the positive supply rail, so the follower characteristics of the *Figure 6.5* circuit are virtually identical to those of *Figure 6.1*, except that it cannot quite follow signals down to zero volts.

Note that the signals appearing on Q_1 collector are amplified and phase-inverted versions of those appearing on the output of the 3140, so the slew rate of the compound follower is typically ten times greater (100 V/μs) than that of the *Figure 6.1* circuit. If care is not taken in the layout, this high slew rate can cause instability. This problem can be overcome by increasing the C_1 value to 1n0, as shown.

So far, the *Figure 6.5* circuit may not seem very useful. This situation can be drastically changed, however, by modifying the circuit as shown in *Figure 6.6*.

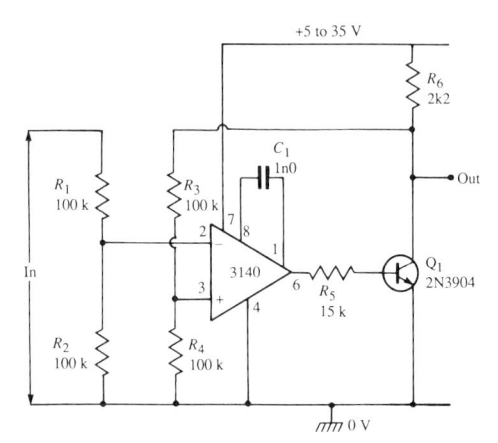

Figure 6.6 *This compound circuit 'follows' input signals to within 50 mV of either supply rail*

Compound circuits 89

Here, the compound op-amp still acts as a non-inverting amplifier, but is given a gain of ×2 via R_3–R_4, and the input signal is attenuated by a factor of two via R_1–R_2, so the circuit still acts (overall) as a unity-gain voltage follower. Note in this case, however, that the input to the 3140 is at only half-supply volts when the input and output of the complete circuit are at full supply voltage value, so this version of the circuit can follow input signals to within 50 mV of either supply rail and has a high slew rate. This circuit is thus greatly superior to the basic *Figure 6.1* design.

An even more dramatic improvement of circuit performance is available from *Figure 6.7*. Here, the 3140 is powered from a 30 V supply and Q_1 is powered from a 50 V supply, so this compound follower can accurately follow input signals with peak values of up to 50 V.

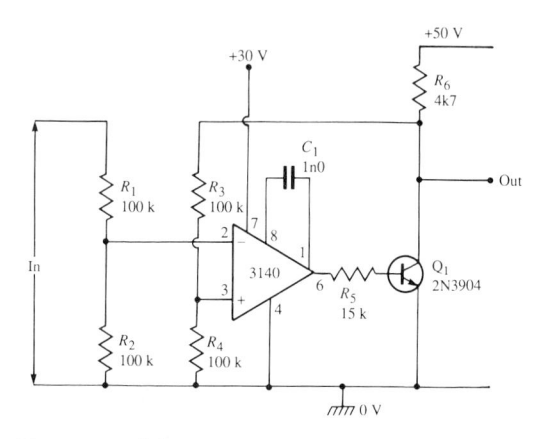

Figure 6.7 *This version of the circuit can follow input signals of up to +50 V peak*

Figure 6.8 shows how the above circuit can be further modified to make a 0–50 V, 0–1 A regulated DC supply. In this case the output of Q_1 is buffered via Darlington emitter follower Q_2–Q_3, which can supply output currents of an amp or so, and the feedback loop to the 3140 is taken from Q_3 emitter, rather than from Q_1 collector. Q_1–Q_2–Q_3 are powered via an unregulated 60 V supply, and the 3140 is powered via a zener-regulated 33 V rail. The compound op-amp is configured as a ×2 non-inverting amplifier, with its input derived from RV_1 slider, which is fully variable from 0 to 25 V; the regulated output of the circuit is thus fully variable over the range 0 to 50 V via RV_1, and output currents of up to an amp or so are available.

Finally, *Figure 6.9* shows how the output stage of the above circuit can be modified to incorporate 1 A overload protection (limiting). R_9 monitors the circuits output current, and when this exceeds 1 A the resulting R_9 volt drop

90 Compound circuits

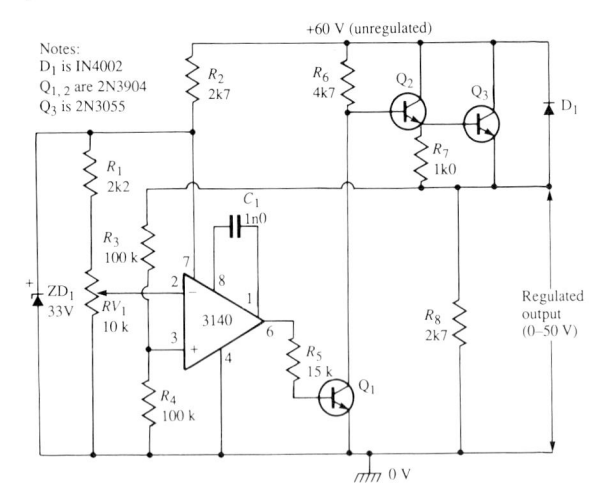

Figure 6.8 *0–50 V, 0–1 A regulated dc supply*

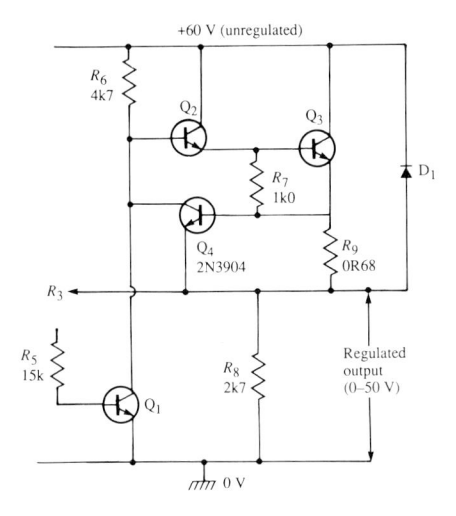

Figure 6.9 *Modified output stage, giving 1 A current limiting to the* Figure 6.8 *circuit*

biases Q_4 on and causes it to rob base-drive current from Q_2–Q_3, thereby limiting the available output current. Note that the feedback connection of R_3 is taken from the R_8–R_9 junction.

Compound circuits 91

Compound inverting amplifiers

Figure 6.10 shows the circuit of a standard op-amp × 100 inverting ac amplifier, operated from a single-ended supply. The circuit's output is biased at half-supply volts (for maximum undistorted signal swing) via potential divider R_3–R_4, and the signal gain is set at × 100 by the R_2/R_1 ratio; the input impedance equals R_1 (10 k), and the op-amp's output can swing to within a few tens of millivolts of the zero voltage rail and to within about 2 V of the positive rail.

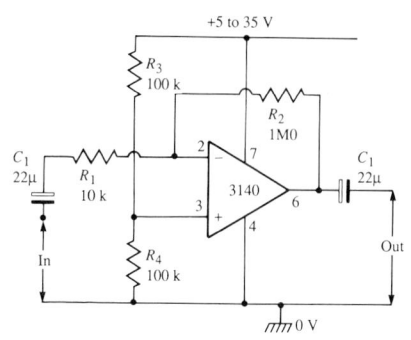

Figure 6.10 *Standard × 100 inverting amplifier op-amp circuit, operated from single-ended supply*

Figure 6.11 shows how to make a compound version of the above amplifier by adding the Q_1 inverting stage and transposing the input connections of the op-amp, as already described for the case of the compound voltage follower.

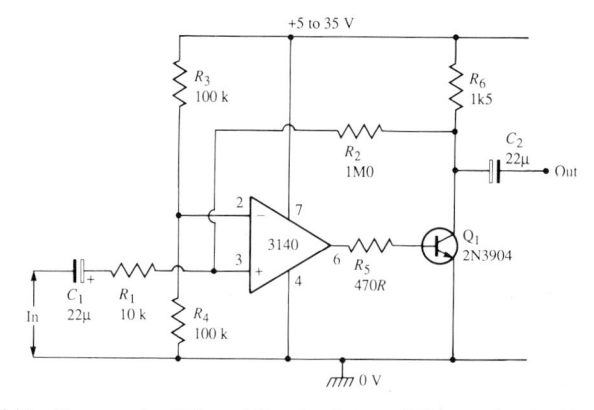

Figure 6.11 *Compound × 100 amplifier circuit has a full-power bandwidth of several hundred kHz. Output can swing within 50 mV of either supply rail*

92 Compound circuits

The gain of this circuit is again set by the R_2/R_1 ratio, and the output (Q_1 collector) is set at half-supply volts via R_3/R_4.

The *Figure 6.11* circuit operates in the same theoretical manner as *Figure 6.10*, except that its output voltage swing is limited by Q_1 rather than by the op-amp, and in practice can swing to within 50 mV of either supply rail before clipping occurs. Also, the slew rate and full-power bandwidth is some ten times higher than in the case of *Figure 6.10*, so this compound design is very superior to the basic 3140 circuit. Because of its high slew rate, the circuit may become unstable if the input signal has a source impedance greater than 2k2.

The available output current (and power) of the *Figure 6.11* circuit is limited to a few tens of milliamperes by R_6, but can easily be increased to any desired value by adding a power-amplifier stage to Q_1 and incorporating it in the amplifier's feedback loop, as shown in *Figure 6.12*. If desired, the simple $D_1-D_2-Q_2-Q_3-R_7-R_8$ amplifier can be replaced by any standard hi-fi type of output stage, making the circuit suitable for use in audio power amplifier systems, etc.

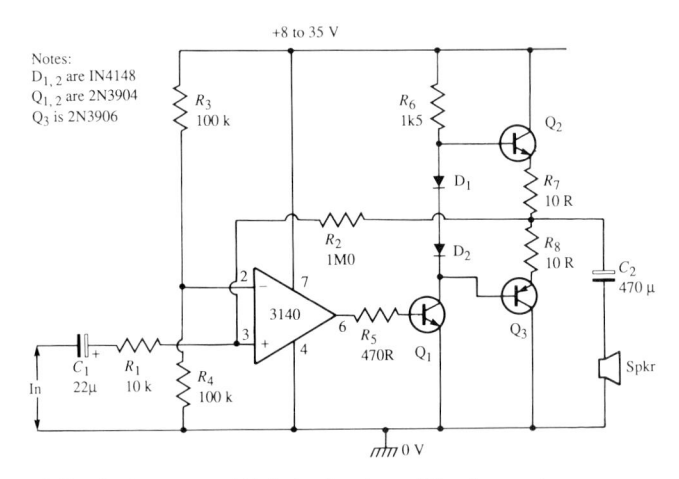

Figure 6.12 *Basic compound hi-fi circuit, with × 100 voltage gain*

The maximum available output voltage swings of the *Figure 6.11* and *6.12* circuits are restricted to 35 V by the supply voltage limitations of the 3140 op-amp. *Figure 6.13* shows how the available output swing can be increased to 120 V (or any other desired value) by powering the 3140 from a 30 V supply but powering Q_1 from a 120 V rail.

Some care must be taken in biasing the *Figure 6.13* circuit, since the 3140 must (for maximum signal swing) be biased to half of its 30 V supply value, while Q_1 must be biased to half of its 120 V value. In the diagram this is

Compound circuits 93

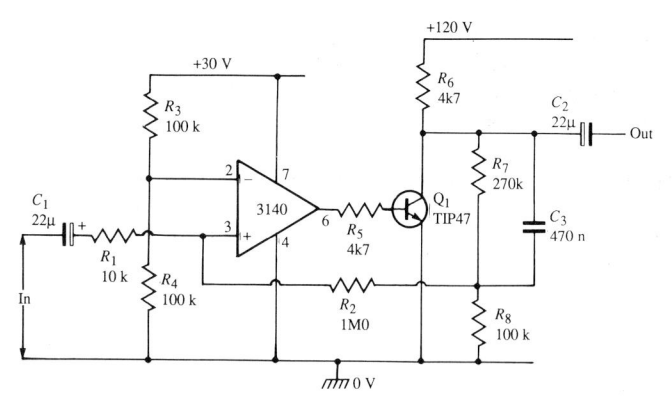

Figure 6.13 *This compound × 100 inverting amplifier gives an output of 120 V peak-to-peak*

achieved by biasing pin 2 of the 3140 to 15 V via R_3–R_4 and interposing a 4:1 (approximately) R_7–R_8 dc divider between the output of Q_1 and the input to feedback resistor R_2, so that 15 V appears on pin 3 of the op-amp when Q_1 collector is at a quiescent half-supply value of 60 V. Note that R_7 is decoupled by C_3, so that the R_7–R_8 dc divider has no significant influence on the ac voltage gain (determined by R_1–R_2) of the circuit.

Compound relaxation oscillators

Figure 6.14 shows the basic circuit of the standard dual-supply op-amp relaxation oscillator or square wave generator (fully described in Chapter 4),

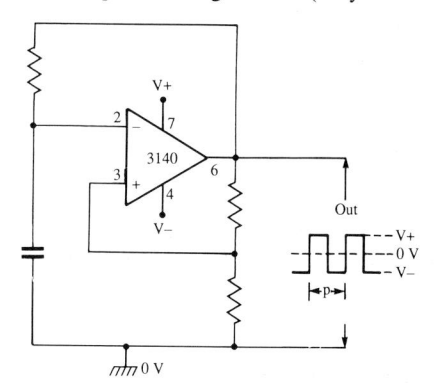

Figure 6.14 *Standard op-amp relaxation oscillator or square wave generator, using a split supply*

94 Compound circuits

and *Figure 6.15* shows it adapted for use with single-ended supplies. The *Figure 6.15* circuit generates a rectangular output waveform which, at any moment of time, is at either zero volts or at about 2 V below the positive supply rail value. The circuit operate as follows.

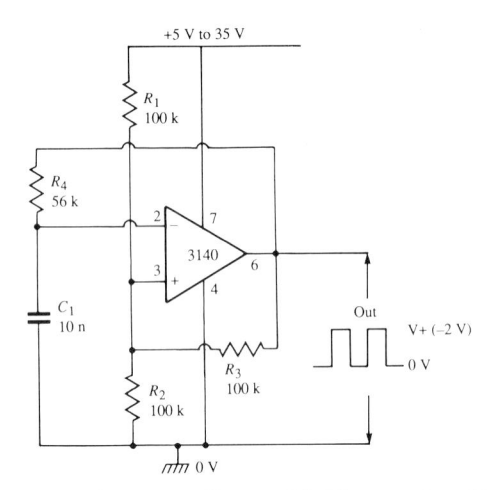

Figure 6.15 *Op-amp relaxation oscillator modified for operation with a single-ended supply*

Suppose that the output has just switched high; in this case R_3 is effectively switched in parallel with R_1, so roughly two thirds of the supply voltage is applied to pin 3, and C_1 charges towards the positive supply rail via R_4 and the op-amp output until the C_1 voltage reaches this pin 3 value, at which point a regenerative switching action is initiated and causes the op-amp output to switch abruptly to zero volts.

Under this new condition R_3 is effectively switched in parallel with R_2, so only one third of the supply voltage is applied to pin 3, and C_1 starts to discharge towards zero via R_4 and the op-amp output until C_1 voltage reaches the new pin 3 value and another regenerative switching action is initiated in which the output switches abruptly high again. The process then repeats *ad infinitum*.

The circuit's waveform period is determined by the R_3, R_4 and C_1 values, and is about 6 ms with the component values shown. The period can be increased (or reduced) by increasing (or reducing) the values of C_1 and/or R_4; C_1 can have any value from 33 pF to 1000 μF, and R_4 can have any value from 10 k to 10 M.

The *Figure 6.15* circuit is quite useful, but suffers from a number of defects. Since its output does not switch to the full positive supply rail voltage when in

Compound circuits 95

the high state, its output waveform is not quite symmetrical, and its period and symmetry vary slightly when the supply voltage is varied. Also, the rise and fall times of the waveform are severely limited by the slew-rate characteristics of the op-amp. When used with a 15 V supply, the output has rise and fall times of 12 μs and 7 μs respectively when feeding a 50 pF load.

Figure 6.16 shows a compound version of the *Figure 6.15* circuit. This is free of all the defects mentioned above; its output waveform is perfectly symmetrical, switches fully between the zero and positive supply rails, has a period that is independent of supply rail values, and has rise and fall times of only 1 μs and 0.7 μs respectively. The circuit is similar to that already described, except for the addition of Q_1–R_5–R_6 and the transposing of the input connections of the 3140 op-amp.

Figure 6.17 shows how the circuit can be modified so that it generates a fixed-frequency rectangular waveform in which the mark-space ratio is fully variable from 25:1 to 1:25 via RV_1. The circuit operation is similar to that already described, except that on high parts of the cycle C_1 charges via R_4–D_2 and the R/H half of RV_1, and on the low parts discharges via R_4–D_1 and the L/H half of RV_1.

2-wire information systems

In the remainder of this chapter we introduce an unusual but very useful concept known as the 2-wire information system, and then show how it can be

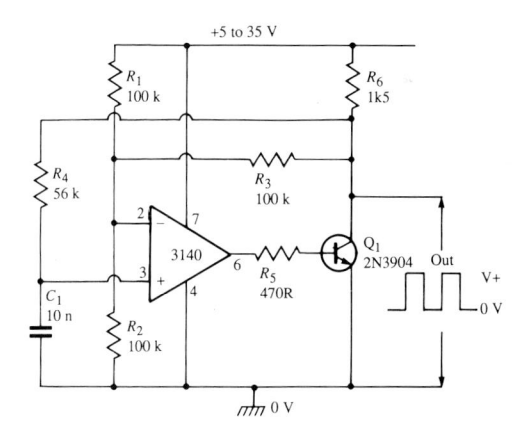

Figure 6.16 *Compound version of the relaxation oscillator. With a 15 V supply, period is 6 ms, rise time 1 μs, and fall time 0.7 μs*

96 Compound circuits

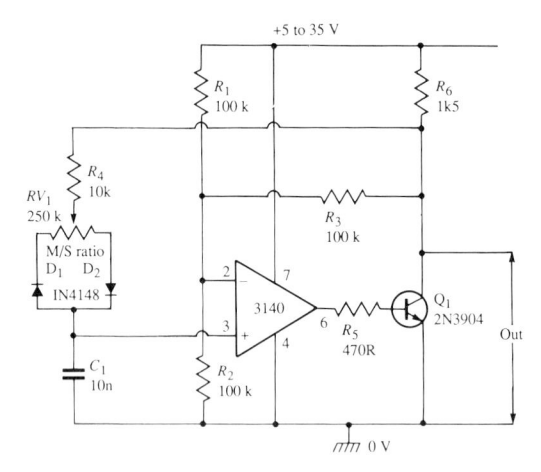

Figure 6.17 *The mark/space ratio of this fixed-frequency compound oscillator is variable from 25:1 to 1:25 via RV_1*

implemented by using compound op-amp voltage comparators. To understand the basic 2-wire concept, compare the circuits of *Figures 6.18* and *6.19*.

Figure 6.18 shows the circuit of a conventional op-amp voltage comparator (fully described in Chapter 3), in which the op-amp output is normally low but switches high when V_{in} rises above V_{ref}, and thus carries information about the relative state of the input voltage. For this circuit to work, it must be powered via a suitable voltage supply, connected via two wires, and its output must be taken to a suitable indicator (a LED or acoustic alarm, etc.) via a third wire, so this is actually a 3-wire information system.

Often, the power supply and output-state indicator (the 'receiver') of a comparator circuit of the *Figure 6.18* type may be located at a base point or

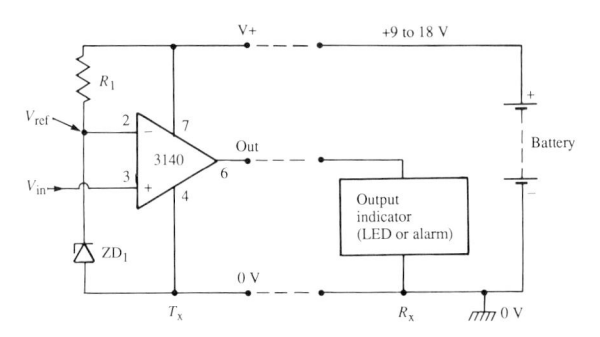

Figure 6.18 *A conventional op-amp comparator acts as a 3-wire information system*

Compound circuits 97

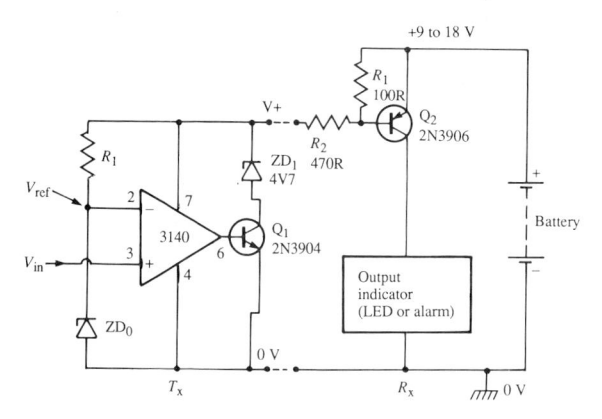

Figure 6.19 *This compound comparator acts as a 2-wire information system*

station, and the actual op-amp and its associated circuitry (the transmitter) may be located at a remote point. If the transmitter and receiver are spaced a significant distance apart, the cost of the 3-wire interconnecting cable may greatly exceed the total cost of the electronic circuitry.

Figure 6.19 shows how a compound op-amp voltage comparator can be used to implement a 2-wire information system by using its supply leads to carry state information, thus reducing cabling costs. In this case the transmitter (the compound voltage comparator) is designed so that it draws either a low current (less than 3 mA) or a high current (greater than 8 mA), depending on the relative states of its input voltages, and draws its supply current from the receiver unit, which acts as a current-monitoring indicator unit. The transmitter supply current flows via R_1–R_2 of the receiver, in which Q_2's base-emitter junction is wired across R_1 and thus activates the output indicator when the transmitter current exceeds 6 mA. The transmitter unit functions as follows.

When V_{in} of the transmitter is below V_{ref} the output of the 3140 is zero, so Q_1 is cut off. Under this condition the op-amp consumes only 2 or 3 mA, and this is not enough to activate the receiver output indicator. When V_{in} is above V_{ref}, however, the 3140 output switches high and turns Q_1 on, making it draw a high current via ZD_1 and R_1–R_2 of the receiver. Under this condition the 3140 supply is pulled down to 4V7 via ZD_1, and the transmitter supply current rises to between 8 and 25 mA (depending on the receiver supply voltage value), thus activating the receiver output indicator.

Thus, the positive transmitter supply rail of this 2-wire system also carries the circuit's 'state' information, which is decoded by the receiver. Note that for correct circuit operation, the minimum (4V7) supply voltage of this compound op-amp circuit must be at least 2 V greater than V_{ref}.

Practical transmitter circuits

In most practical applications of the *Figure 6.19* 2-wire system the inputs to the voltage comparators are obtained from a Wheatstone bridge network in which one of the elements is a resistive transducer sensitive to light or heat, etc. *Figures 6.20* to *6.24* show five variations of this type of transmitter. In all of these circuits a half-supply reference voltage is applied to one input pin of the op-amp via R_1–R_2, and a variable voltage is applied to the other. Since these voltages are bridge-derived, the balance or trigger points of the circuits are independent of the op-amp supply rail values and are determined only by the resistance ratios of the input bridge.

Figures 6.20 and *6.21* show light-sensitive transmitter circuits in which a cadmium-sulphide photocell or light-sensitive resistor (LDR) is used as the sensing element. The LDR and RV_1 should have nominal values of at least 10 k. In *Figure 6.20* the LDR is wired above RV_1, and the pin-3 voltage thus rises as the light intensity increases and the LDR resistance falls, and the

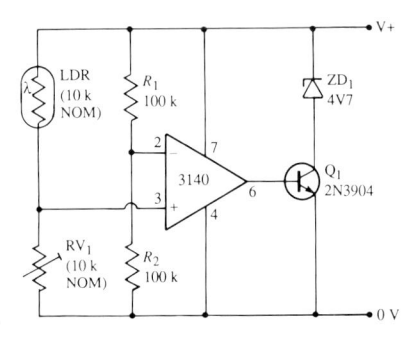

Figure 6.20 *Light-activated 2-wire transmitter*

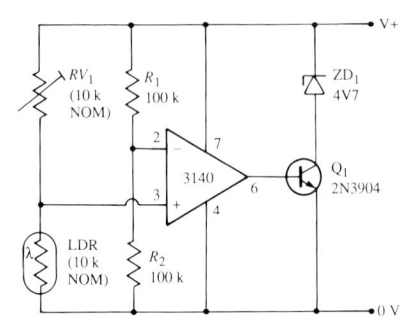

Figure 6.21 *Dark-activated 2-wire transmitter*

Compound circuits 99

circuit thus acts as a 'light-activated' transmitter. In *Figure 6.21* the LDR is wired below RV_1, so the pin-3 voltage rises as the light intensity falls and the LDR resistance increases, and this circuit thus acts as a 'dark-activated' transmitter.

Figures 6.22 and *6.23* show how the above circuits can be modified for use as temperature-sensitive transmitters by using a NTC thermistor (nominal value 10 k) in place of the LDR. The output of the *Figure 6.22* circuit switches high (draws a heavy current) when the TH_1 temperature exceeds a value pre-set via RV_1, and that of *Figure 6.23* switches high when the temperature falls below a value pre-set via RV_1.

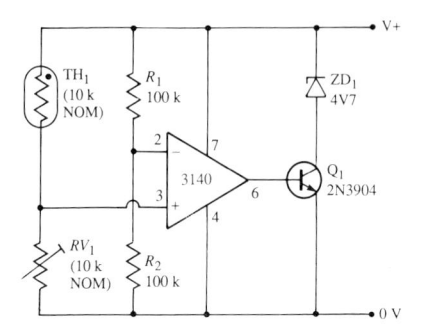

Figure 6.22 *Over-temperature 2-wire transmitter*

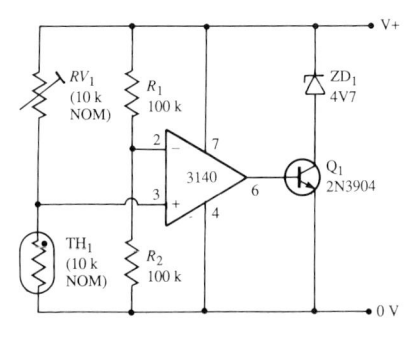

Figure 6.23 *Under-temperature 2-wire transmitter*

Figure 6.24 shows the circuit of a 2-wire transmitter that gives a high output when the level of a liquid exceeds a pre-set value. In this case the 3140 pin connections are transposed relative to the earlier circuits. The ground connection is taken to the liquid that is being monitored, and the tip of the metal probe is placed at the required alarm level in the liquid container. When

100 Compound circuits

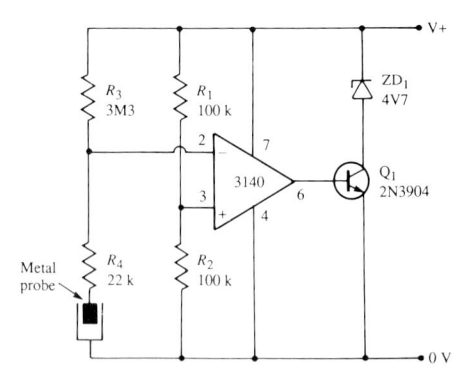

Figure 6.24 *Liquid-level 2-wire transmitter*

the liquid is below the probe tip, pin 2 of the op-amp output is low. When the liquid reaches the probe, its resistance pulls pin 2 below pin 3, and the op-amp output switches high. The circuit action can be reversed, so that the output goes high when the liquid falls below a pre-set level, by simply transposing the pin 2 and 3 connections. With the R_3 value shown, the liquid resistance must be below 3.3 MΩ for correct operation.

A 2-wire receiver

Finally, to complete this chapter, *Figure 6.25* shows the circuit of a 2-wire receiver that incorporates an audio-visual output indicator and which can be

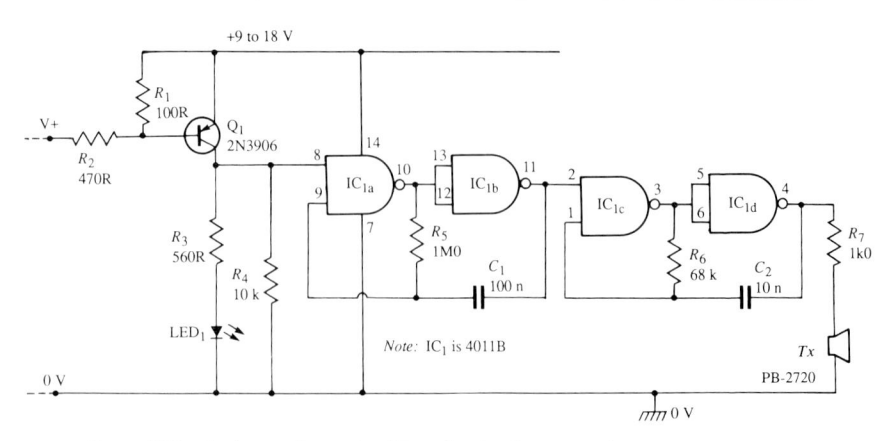

Figure 6.25 *2-wire receiver unit with audio-visual output indication*

used with any of the *Figure 6.19* to *6.24* circuits. When a high output is detected from the transmitter, Q_1 turns on and its collector is pulled high, simultaneously driving LED_1 on via R_3 and activating gated double astable multi-vibrator IC_1, which produces a pulsed-tone signal in Tx, which is a low-cost high-efficiency PB2720 or similar type of acoustic transducer.

7 Norton op-amp circuits

The most popular op-amps, such as the 741, CA3140, LF351, etc., give an output proportional to the difference between their two input pin voltages (*Figure 7.1(a)*), and are thus known as voltage-differencing amplifiers (VDAs). There is, however, a type of op-amp which gives an output voltage proportional to the difference between the currents applied to its two input pins, and this is known as a current-differencing amplifier (CDA). *Figure 7.1(b)* shows the standard symbol of a CDA, which is also known as a Norton op-amp.

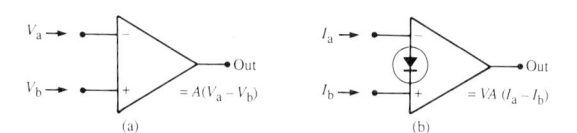

Figure 7.1 *A conventional op-amp (a) is a voltage-differencing amplifier, but a Norton op-amp (b) is a current-differencing device*

The two best known versions of the Norton op-amp are the LM3900 and the LM359. The LM3900 is a low-cost medium performance IC that houses four identical op-amps in a 14-pin DIL package (see *Figure 7.2*) and can operate from a single-ended 4 to 36 V power supply. Each of its op-amps has a 2.5 MHz unity-gain bandwidth and a 70 dB open-loop gain, and gives a large output voltage swing. This IC is very useful in dc and low-frequency applications where several op-amp stages are needed in single-ended-supply circuits.

The LM359, on the other hand, is a very fast dual Norton amplifier in which each op-amp has a 30 MHz unity-gain bandwidth and a 72 dB open-loop gain, and in which most of the op-amp parameters are externally

Norton op-amp circuits 103

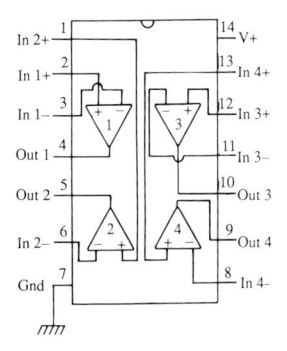

Figure 7.2 *Connections of the LM3900 quad Norton op-amp*

programmable. This IC is particularly useful in video and high-frequency amplifier/filter applications.

The LM3900 and LM359 operate in a very different way to conventional op-amps, and require the use of special biasing techniques. This chapter explains how the devices work, and how to use them in a variety of practical applications.

LM3900 basic principles

The LM3900 incorporates four identical current-differencing op-amps, each having the circuit shown in *Figure 7.3*. To aid understanding of the complete

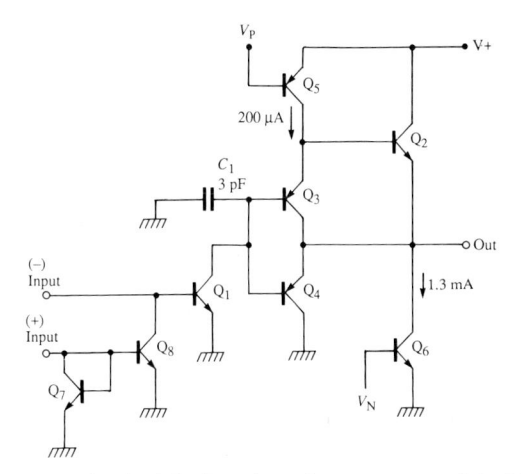

Figure 7.3 *Circuit of each of the four identical op-amp stages of the LM3900*

circuit, *Figure 7.4* shows six simple stages in the development of the final design.

Figure 7.4(a) shows a basic inverting amplifier circuit. Q_1 is a common emitter amplifier with a high-impedance (constant-current) collector load, and gives a high-gain inverting action, and Q_2 is an emitter follower output buffer with a high-impedance emitter load. The high frequency response of the complete amplifier is rolled off by C_1, to enhance circuit stability. Note that the output of this circuit can swing to within a few hundred millivolts of both zero and the positive supply rail voltage, and that the overall current gain of the circuit equals the product of its two transistor current gains.

Figure 7.4(b) shows how the current gain of the above circuit can be increased with little loss of available output voltage swing, by adding pnp transistor Q_3. The output of this circuit can typically source up to 10 mA (via Q_2), but can sink only 1.3 mA (via Q_2's constant-current generator). *Figure 7.4(c)* shows how the sink current can be increased under overdrive conditions by wiring Q_4 so that it gives class-B operation under the overdrive condition.

Figure 7.4(d) shows the appearance of the above circuit when transistors Q_5 and Q_6 are used as constant-current generators. These two generators are biased via a network that is built into the LM3900 IC.

Current mirrors

The *Figure 7.4(d)* circuit forms the basis of each of the LM3900 amplifier stages, but gives an inverting action only. The non-inverting action of the LM3900 is provided with the assistance of the current mirror circuit of *Figure 7.4(e)*, which is made up of two matched and integrated transistors and simply draws an output current that is almost identical to the input drive current. The circuit operates as follows.

The input current of the *Figure 7.4(e)* circuit is fed to the bases of both transistors. Suppose these have current gains of × 100, and both transistors are drawing identical base currents of 5 μA. In this case both transistors draw collector currents of 500 μA. Note, however that the Q_7 collector current is drawn from the circuit's input current, which thus equals 500 μA plus (2 × 5 μA), or 510 μA, and that the Q_8 collector current is the output or mirror current of the circuit. The input and output currents of the circuit are thus almost identical (within a few per cent), irrespective of the input current magnitude.

Figure 7.4(f) shows how the current mirror circuit can be connected to the basic *Figure 7.4(a)* circuit to give the current-differencing action of the Norton amplifier. Here, the mirror circuit is driven via the non-inverting

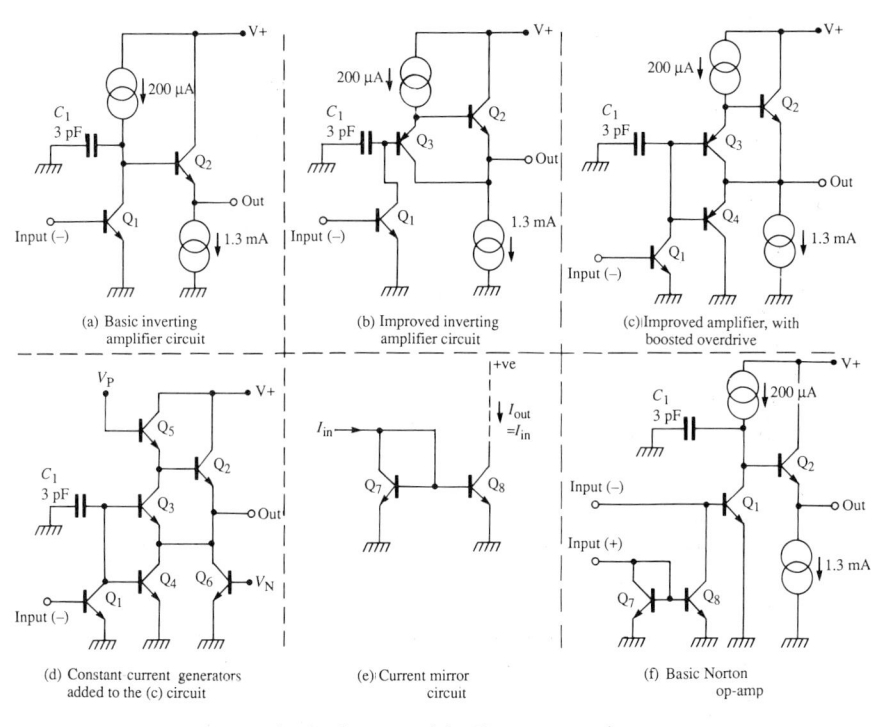

Figure 7.4 *Stages in the development of the* Figure 7.3 *circuit*

(a) Basic inverting amplifier circuit
(b) Improved inverting amplifier circuit
(c) Improved amplifier, with boosted overdrive
(d) Constant current generators added to the (c) circuit
(e) Current mirror circuit
(f) Basic Norton op-amp

input terminal, and the mirror current is drawn from the inverting input terminal, which is also connected directly to the base of the Q_1 amplifier stage. Consequently, the base current of Q_1 is equal to $(I-)-(I+)$, and is thus equal to the difference between the two input currents. The complete amplifier (*Figure 7.3*) thus gives the current-differencing op-amp action already mentioned.

Note that since both input terminals of the op-amp are connected to transistor base-emitter junctions, both inputs act (in voltage terms) as virtual-ground points. Consequently, these CDA circuits can be made to act like conventional voltage-differencing op-amps by wiring high-value resistors in series with their input terminals, so that the input currents are directly proportional to the input voltage/resistor values; when this technique is used, there is no upper limit to the available input common-mode voltage range of the LM3900 op-amp.

Biasing techniques

The basic amplifier stages of the LM3900 have high current gains, and the output of the amplifier starts to swing down through the half-supply point when the input bias current of Q_1 starts to rise above 30 nA or so. This input current is normally equal to the difference between the two input terminal currents; which should normally be restricted to the range 0.5 μA to 500 μA (ideally about 10 μA).

In linear applications an op-amp is normally biased so that its output takes up a quiescent value of half-supply volts, to accommodate maximum undistorted signal swings, and *Figure 7.5(a)* shows how the LM3900 can be biased to meet this condition. R_1–R_2–C_1 generate a decoupled half-supply reference voltage, which applies a reference current to the non-inverting terminal via R_3, and a negative feedback current is applied from the op-amp output to the inverting terminal via R_4. The basic action is such that the op-amp output automatically adjusts to such a value that the two input currents equalize and hence reduce the internal Q_1 base current to near-zero (about 30 nA), and in the case of *Figure 7.5(a)* this situation occurs when V_{out} equals V_{ref}. In practice, the single reference voltage source can be used to apply biasing to several op-amp stages.

A variation of this biasing system is shown in *Figure 7.5(b)*. In this case the non-inverting terminal is biased from the positive supply rail via R_1, which has a value approximately double that of R_2, causing the output to bias at a quiescent value of half-supply volts. A minor defect of this biasing technique is that it allows supply line ripple to break through to the output, with a gain of $\times 0.5$.

Note in the *Figure 7.5(a)* and *(b)* circuits that the input signal is shown connected to the inverting terminal of the amplifier but that in practice the signal can alternatively be connected to the non-inverting input.

Finally, *Figure 7.5(c)* shows an alternative biasing technique that can be used when the op-amp is to be operated only as an inverting amplifier. In this case the non-inverting terminal is disabled, and feedback potential divider R_1–R_2 is applied between the output and the inverting terminal. Consequently, since the inverting terminal acts as a transistor base-emitter junction (with a V_{be} value of about 0.55 V at 10 μA bias), the output automatically takes up a quiescent value of $V_{be} \times (1 + R_1/R_2)$, or about 6 V with the component values shown.

Linear amplifier circuits

Figures 7.6 to *7.11* show six ways of using LM3900 op-amps as linear amplifiers. In the *Figure 7.6* circuit R_2 and R_3 bias the output to a quiescent

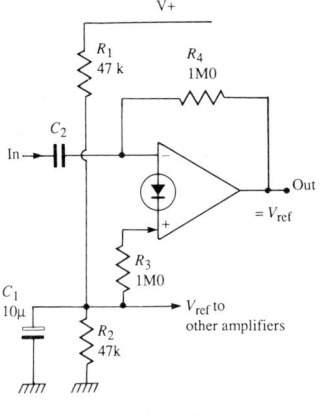

(a) Voltage-reference biasing

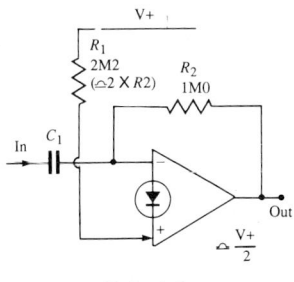

(b) Supply-line biasing

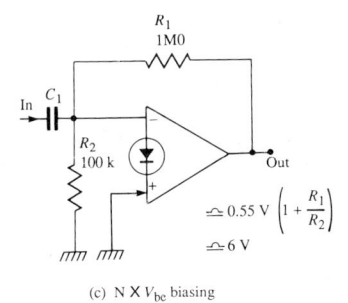

(c) $N \times V_{be}$ biasing

Figure 7.5 *Methods of biasing LM3900 op-amps for linear operation*

108   Norton op-amp circuits

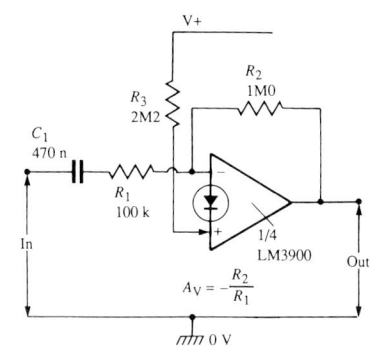

Figure 7.6 *Inverting ac amplifier with supply-line biasing*

half-supply value, using the technique shown in *Figure 7.5(b)*. The input signal is applied to the inverting terminal via R_1, and the voltage gain is determined by the R_1–R_2 ratio, so this design acts as a ×10 inverting amplifier. *Figure 7.7* shows an alternative version of the ×10 inverting amplifier, in which N × V_{be} biasing is used and the gain is determined by the R_1–R_2 ratio.

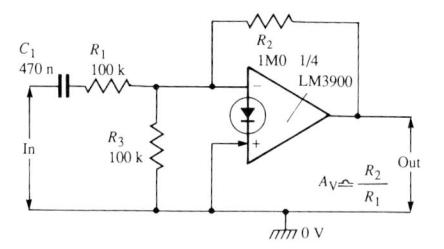

Figure 7.7 *Inverting ac amplifier with N × V_{be} biasing*

Figure 7.8 shows the connections for making a non-inverting amplifier with a gain of approximately ×10. Supply-rail biasing is again used, but the input signal is applied to the non-inverting pin via R_1.

The LM3900 op-amps are fairly slow devices; they have slew rates of only 0.5 V/µs, and thus have very limited useful bandwidths. *Figure 7.9* shows how the useful bandwidth can be increased by connecting an external common emitter transistor to the output and transposing the input connections of the standard amplifier to make a ×100 compound amplifier with a 200 kHz bandwidth. Because of its very high overall gain, this circuit may be unstable if

Norton op-amp circuits 109

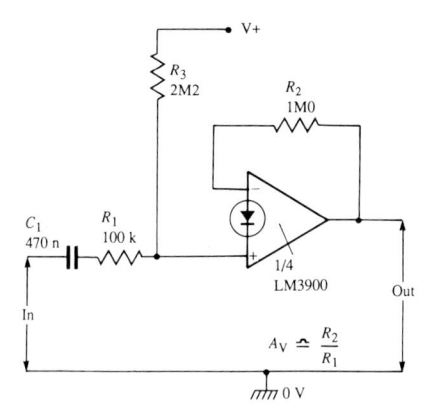

Figure 7.8 *Non-inverting amplifier*

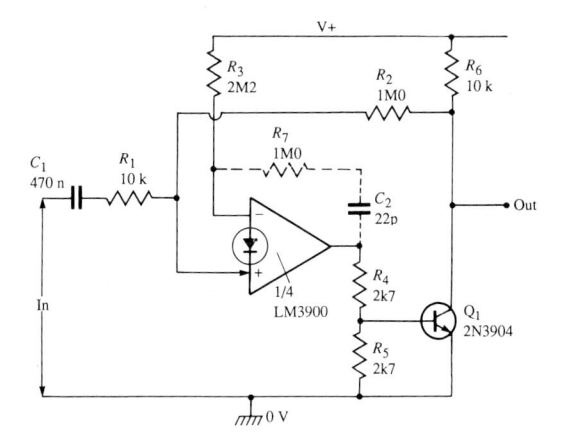

Figure 7.9 *Wideband (200 kHz) high-gain (× 100) amplifier*

care is not taken in layout. R_7 and C_2 can be used to slightly reduce the bandwidth and enhance stability if required.

Figure 7.10 shows how the above circuit can be modified to give a peak-to-peak output voltage swing of 150 V (or whatever voltage is used to power Q_1). Note that the output voltage of this circuit has a quiescent value of 75 V, causing 7.5 µA to be fed to the non-inverting terminal of the op-amp via R_2, so, to give correct biasing, R_3 (powered from the 15 V supply rail of the op-amp) must also apply 7.5 µA to the inverting pin of the op-amp, as shown.

Finally, *Figure 7.11* shows how to connect an LM3900 op-amp as a unity-gain non-inverting amplifier or voltage following buffer. The input is

110 Norton op-amp circuits

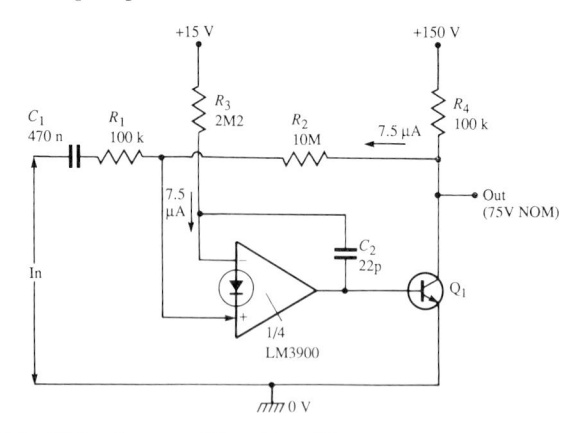

Figure 7.10 *High-voltage amplifier with ×100 gain*

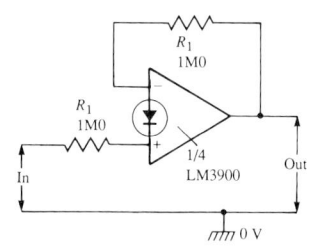

Figure 7.11 *dc voltage-following buffer*

connected to the non-inverting terminal via R_1, thus giving the non-inverting action, and R_1 and R_2 have equal values, thus giving unity gain (note that the circuit would give a gain of ×2 if R_1 were half the value of R_2).

Comparators and schmitt circuits

The LM3900 op-amp can be made to act as a voltage comparator by simply wiring equal value current limiting resistors in series with each input, and then using one resistor as the input point of the voltage reference and the other as the sample input point, as shown in the circuits of *Figures 7.12* to *7.14*. The *Figure 7.12* circuit gives inverting voltage comparator action, in which the output switches high when V_{in} falls below V_{ref}, and the *Figure 7.13* circuit gives non-inverting voltage comparator action, in which the output switches high when V_{in} rises above V_{ref}.

Norton op-amp circuits 111

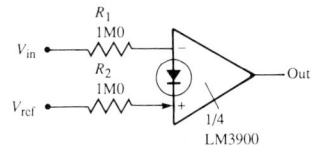

Figure 7.12 Inverting voltage comparator

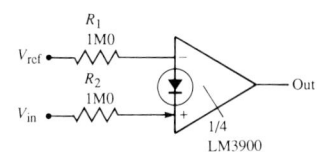

Figure 7.13 Non-inverting voltage comparator

The *Figure 7.12* and *7.13* comparator circuits can supply output currents of only a few milliamperes. The available output current can be boosted to tens or hundreds of milliamperes by connecting a common emitter transistor stage to its output, as shown in the non-inverting power comparator circuit of *Figure 7.14*.

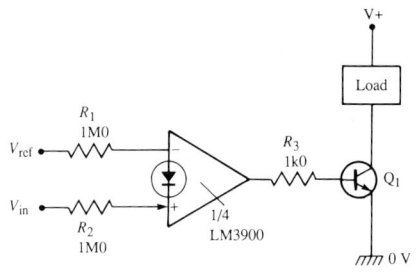

Figure 7.14 Non-inverting power comparator

Hysteresis can be added to the LM3900 voltage comparator circuits, so that they act as Schmitt triggers, by simply connecting a high-value resistor between the output and the non-inverting terminal, as shown in *Figures 7.15*

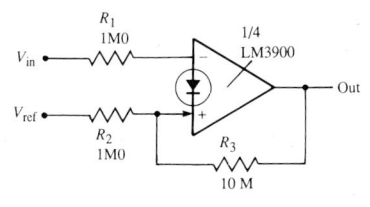

Figure 7.15 Inverting Schmitt trigger

112   Norton op-amp circuits

and 7.16. The *Figure 7.15* circuit gives an inverting Schmitt action, and *Figure 7.16* gives a non-inverting Schmitt action. The R_2–R_3 ratio determines the hysteresis magnitude.

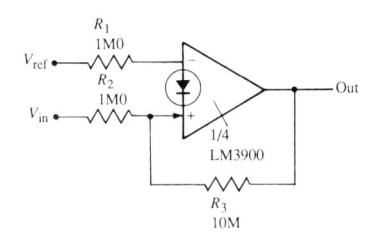

Figure 7.16   *Non-inverting Schmitt trigger*

Comparator applications

Figures 7.17 to *7.21* show some useful applications of voltage comparators. The *Figure 7.17* design is that of an over-temperature switch, the output of which goes high when the temperature of NTC thermistor TH_1 exceeds a value pre-set via RV_1. Potential divider R_1–R_2 feeds a fixed half-supply reference voltage to R_3, which then feeds a reference current to the inverting terminal, and TH_1–RV_1 form a potential divider that feeds a variable current to the non-inverting input via R_4. The potential on the TH_1–RV_1 junction rises with temperature, and the op-amp output switches high when this voltage exceeds half-supply value; the trip temperature can be pre-set via RV_1.

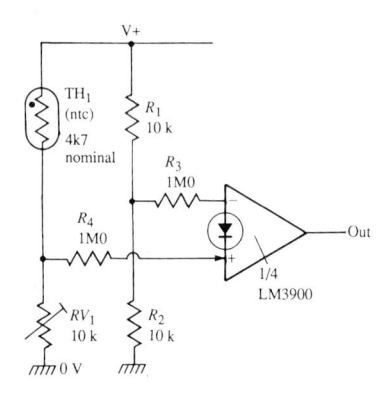

Figure 7.17   *Over-temperature switch*

Norton op-amp circuits 113

Note that the operation of the above circuit can be reversed, so that it operates as an under-temperature switch, by transposing the TH_1–RV_1 positions. Also note that, since RV_1–TH_1–R_1–R_2 are wired in a Wheatstone bridge configuration, the trip point is independent of supply rail variations.

Figure 7.18 shows a useful variation of the above circuit, wired as an under-temperature switch. In this case the reference (inverting) current is derived from the supply rail via R_1, and the variable (non-inverting) current is again derived from the RV_1–TH_1 junction. Since the R_1 value is roughly double that of R_2 and generates a current proportional to the supply rail voltage, the trip point of this circuit is also independent of variations in supply rail voltage.

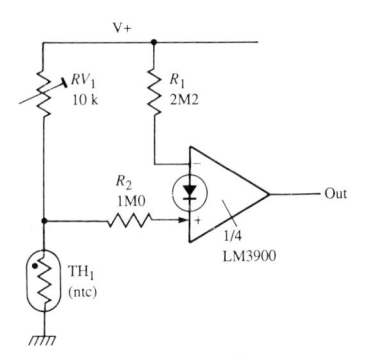

Figure 7.18 Under-temperature switch

A variant of the above is shown in *Figure 7.19*, which gives a high output when the supply voltage falls below a value determined by ZD_1. If ZD_1 has a value of 5V6, the op-amp output switches high when the supply rail voltage falls below roughly 11 V; the precise trip point can be varied by replacing R_3 with a series-connected 820 k resistor and a 470 k pot.

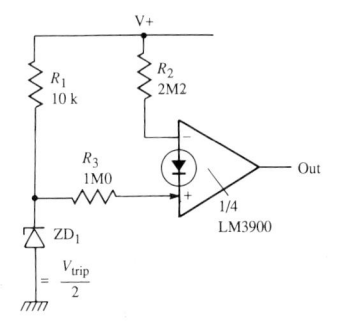

Figure 7.19 Supply under-voltage detector

114 Norton op-amp circuits

Finally, *Figures 7.20* and *7.21* show how the comparator can be made to act as a 3-input logic gate. In *Figure 7.20*, a reference current is fed to the inverting pin via R_4, and a greater current can be fed to the non-inverting pin via any of the R_1 to R_3 resistors, thus causing the output to switch high if any of the input terminals go high; this circuit thus acts as a 3-input OR gate.

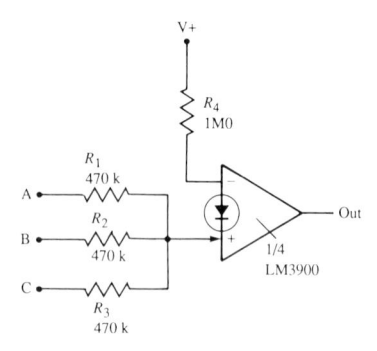

Figure 7.20 3-input OR gate (can be converted to a NOR gate by transposing the op-amp inputs)

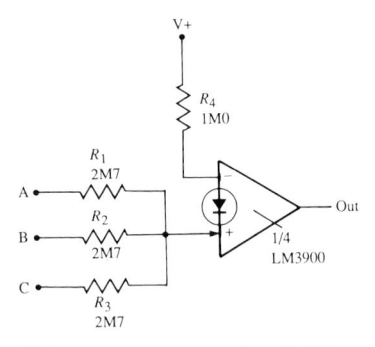

Figure 7.21 3-input AND gate (can be converted to a NAND gate by transposing the op-amp inputs)

Note that this circuit can be made to accept any desired number of OR inputs by simply using a suitable number of input resistors, and that the circuit can be converted into a NOR gate by simply transposing the input connections of the op-amp.

The *Figure 7.21* circuit is that of a 3-input AND gate, which gives a high output only when all three inputs are taken high, making the non-inverting input terminal current exceed that of R_4. This circuit can be converted to a NAND gate by transposing the input connections of the op-amp.

Norton op-amp circuits 115

Voltage regulator circuits

Figures 7.22 to *7.26* show various ways of using LM3900 op-amps as simple voltage regulators and references. The *Figure 7.22* circuit is a simple but useful variable voltage reference. The non-inverting terminal of the op-amp is disabled, and the circuit uses the V_{be} potential of the inverting terminal as a reference, and has a voltage gain determined by the RV_1–R_1 ratio. When RV_1 is set to zero, the circuit gives unity gain and gives an output of 0.55 V; when RV_1 is set to the maximum value the circuit has a gain of × 50 and gives an output of 25 V. The circuit has good regulation and can supply output currents of several milliamperes; note, however, that the output voltage is not temperature compensated.

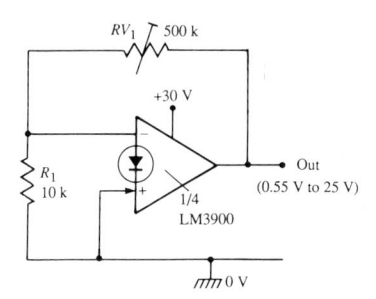

Figure 7.22 *Simple variable-voltage reference*

Figure 7.23 shows a fixed voltage reference circuit that generates a well regulated output that is slightly greater than the ZD_1 voltage. R_1 sets the zener current at about 1 mA. The circuit can safely supply output currents of only a few milliamperes, but this can easily be boosted to tens or hundreds of milliamperes by wiring a current boosting transistor into the output feedback loop of the circuit, as shown in *Figure 7.24*.

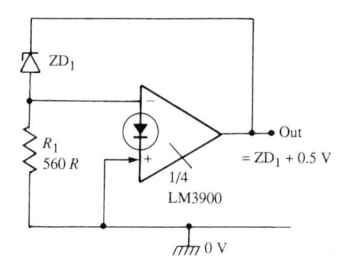

Figure 7.23 *Fixed-voltage reference*

116 Norton op-amp circuits

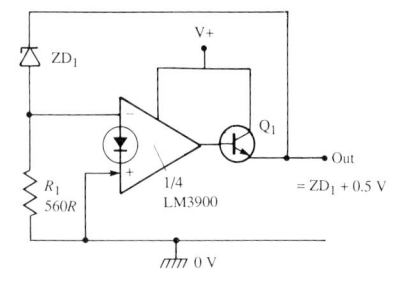

Figure 7.24 *Current-boosted voltage reference*

Figure 7.25 shows an alternative type of voltage regulator, which gives a well regulated variable voltage output. In this case the op-amp is wired as a ×2 non-inverting dc amplifier (with gain determined by the R_3–R_2 ratio), and the input voltage is variable from zero to 15 V via RV_1; the output voltage is thus variable over the approximate range 0.5 V to 30 V via RV_1. *Figure 7.26* shows how the available current can be boosted to tens or hundreds of milliamperes with the aid of an external transistor.

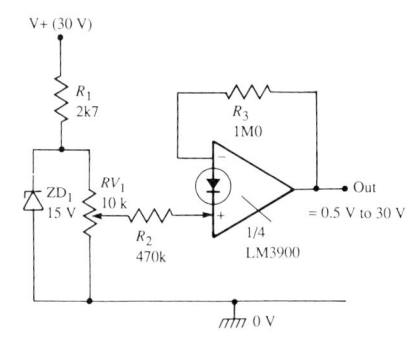

Figure 7.25 *Variable voltage regulator*

Current regulator circuits

Figures 7.27 to *7.30* show various ways of using the LM3900 to make fixed current regulator circuits. The *Figure 7.27* design acts as a fixed (1 mA) current source, which feeds a fixed current into a load connected between Q_1 collector and ground almost irrespective of the load impedance (in the range zero to 14 k). The circuit is powered from a regulated 15 V supply. Potential divider R_1–R_2 applies a 14 V reference (15 V–1 V) to R_3, so the op-amp output automatically adjusts to give an identical voltage at the R_4–R_5 junction. This

Norton op-amp circuits 117

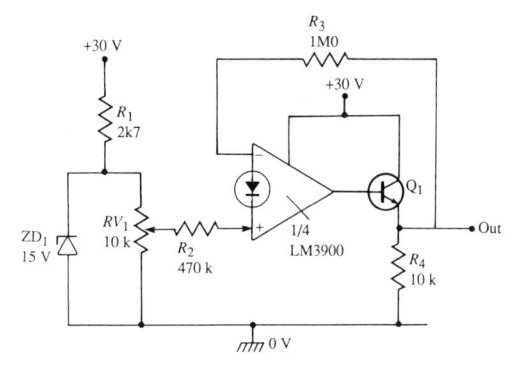

Figure 7.26 *Variable voltage regulator with boosted output*

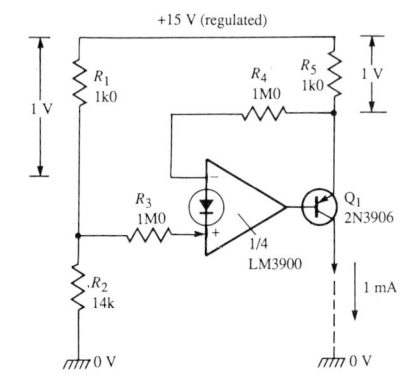

Figure 7.27 *Fixed-current source (1 mA)*

produces 1 V across R_5, resulting in an R_5 current of 1 mA. Since this current is derived from Q_1 emitter, and the emitter and collector currents of a transistor are almost identical, the circuit acts as a fixed current source. The source current can be doubled, if desired, by halving the R_5 value, etc.

Figure 7.28 shows a simple variation of the above circuit, in which the source current is independent of variations in supply rail voltage. In this case the input is set to 2V7 below the supply rail value via ZD_1, so 2V7 is automatically set across R_4, which has a value of 2k7 and thus produces a fixed 1 mA source current from Q_1.

Figure 7.29 shows a simple 1 mA current sink, in which a fixed current flows in any load connected between the positive supply rail and Q_1 collector, almost irrespective of the load impedance. Here, the non-inverting terminal of the op-amp is disabled, and 100% negative feedback is used between the

118   Norton op-amp circuits

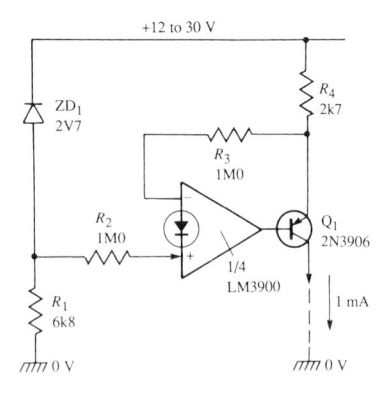

Figure 7.28   *Alternative current source (1 mA)*

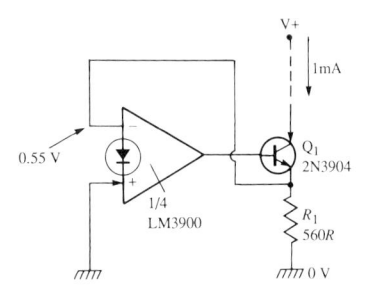

Figure 7.29   *Simple 1 mA current sink*

output of the circuit (Q₁ emitter) and the inverting terminal. The voltage across R_1 thus equals the V_{be} of the inverting terminal and, since this is roughly 0.55 V, a fixed current of about 1 mA flows through Q₁ emitter and R_1, and thus into Q₁ collector from any load that is connected. Note that the sink current of this circuit is not temperature compensated.

Finally, *Figure 7.30* shows an alternative type of current sink. In this case the op-amp is fully enabled, and has a fixed reference of 2V7 applied to its non-inverting terminal via R_2. Consequently, the circuit automatically adjusts to generate 2V7 across R_4 which, since it has a value of 2k7, generates a current of 1 mA in the emitter and collector of Q₁. This current can be varied, if required, either by varying the value of R_4 or by varying the input voltage feed to R_2.

Norton op-amp circuits 119

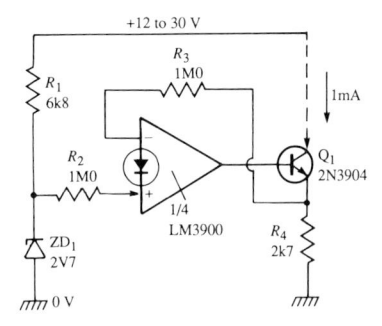

Figure 7.30 *Improved current sink (1 mA)*

Waveform generator circuits

To conclude this, look at the LM3900. *Figures 7.31* to *7.35* show some useful ways of using the op-amps to make simple waveform generators. *Figure 7.31* shows a 1 kHz square wave generator, in which C_1 alternately charges and discharges via R_1. When the output is high, R_3–R_4 are effectively connected in parallel, and C_1 charges via R_1 until the current flow into R_2 equals that flowing into the non-inverting terminal of the op-amp; this point occurs when the voltage across C_1 rises to roughly two thirds of $+V$. At this point the circuit switches regeneratively, the output jumps low, and C_1 starts to discharge via R_1. Under this condition R_4 is effectively disabled and the input current to the non-inverting terminal is determined only by R_3, so C_1 discharges until the R_2 current falls slightly below that of R_3. This point occurs when the C_1 voltage falls to about one third of $+V$. At this point the circuit again switches regeneratively, and the output goes high again. The action then repeats *ad infinitum*.

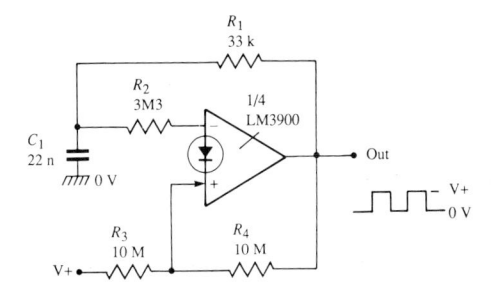

Figure 7.31 *1 kHz square wave generator*

120 Norton op-amp circuits

The *Figure 7.31* circuit is useful for generating square waves with frequencies up to a maximum of only a few kHz; because of the poor slew rate performance of the LM3900 (0.5 V/μs), the output waveform has fairly poor rise and fall times. The circuit generates a symmetrical square wave output. *Figure 7.32* shows how the circuit can be modified to give a variable mark-space (M/S) ratio output. In this case C_1 alternately charges via R_1–D_1 and the upper half of RV_1, and discharges via R_1–D_2 and the lower half of RV_1. The M/S ratio can be varied over the approximate range 1:10 to 10:1 via RV_1.

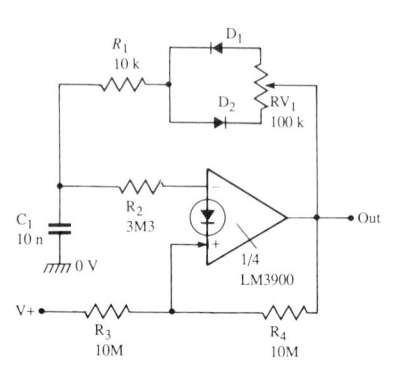

Figure 7.32 *Variable mark/space ratio generator*

Figure 7.33 shows a simple modification of the above circuit, which causes it to act as a free-running pulse generator. In this case C_1 alternately charges via R_1–D_1 and discharges via R_2, producing an M/S ratio of about 1:60.

Figure 7.34 shows how the basic *Figure 7.31* circuit can be modified to act as a gated 1 kHz astable or square wave generator by taking R_3 to ground via R_5, rather than directly to the positive supply rail. The circuit becomes active only when the gate terminal is pulled high (to the positive supply rail).

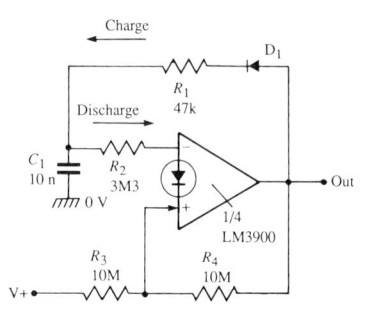

Figure 7.33 *Pulse generator*

Norton op-amp circuits 121

Finally, to complete this look at the LM3900, *Figure 7.35* shows how the *Figure 7.34* and *7.17* circuits can be combined to make an audible-output over-temperature alarm, which generates a 1 kHz tone in the PB-2720 (or similar) acoustic transducer when the TH_1 temperature exceeds a value preset via RV_1.

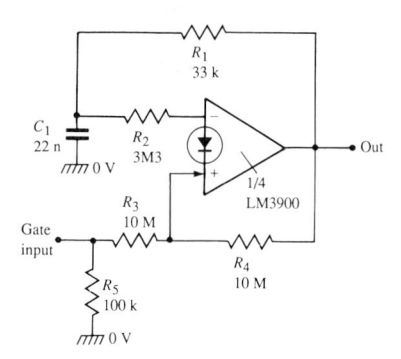

Figure 7.34 *Gated 1 kHz astable*

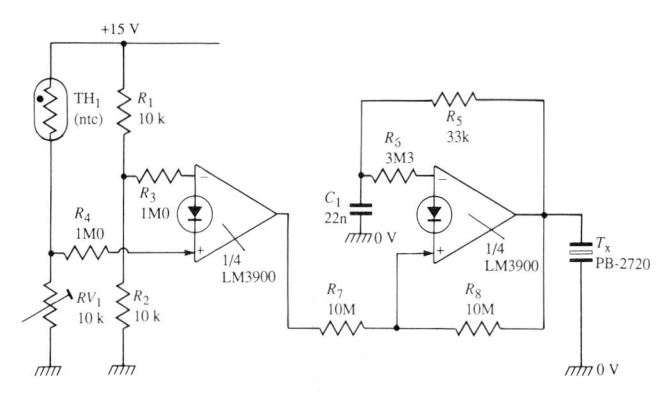

Figure 7.35 *Audible-output over-temperature alarm*

The LM359 dual op-amp

The LM359 is not as well known as the LM3900, but is an outstandingly useful Norton type of IC. The LM359 is in fact a high performance IC that houses two identical Norton op-amps, plus a common biasing network, in a

122   Norton op-amp circuits

14-pin DIL package (see *Figure 7.36*) and can operate from a single-ended 5 V to 22 V power supply. Each of its op-amps offers a 30 MHz unity-gain bandwidth, a 60 V/μs slew rate, and a 72 dB open-loop gain, and has many of its parameters fully programmable via one or two external resistors.

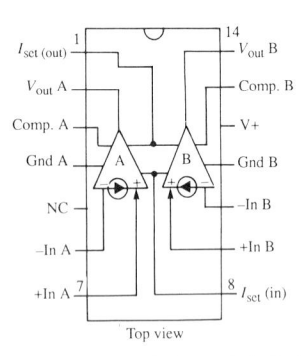

Figure 7.36   *Outline and pin notations of the LM359 dual high-speed programmable Norton amplifier*

The LM359's op-amps differ considerably from those used in the LM3900. *Figure 7.37* shows the basic LM359 op-amp circuit in slightly simplified form. This consists, in essence, of a mirror-driven (via Q_1–Q_2) wide-band cascode amplifier (Q_3 and Q_4), which does not suffer from output-to-input Miller or

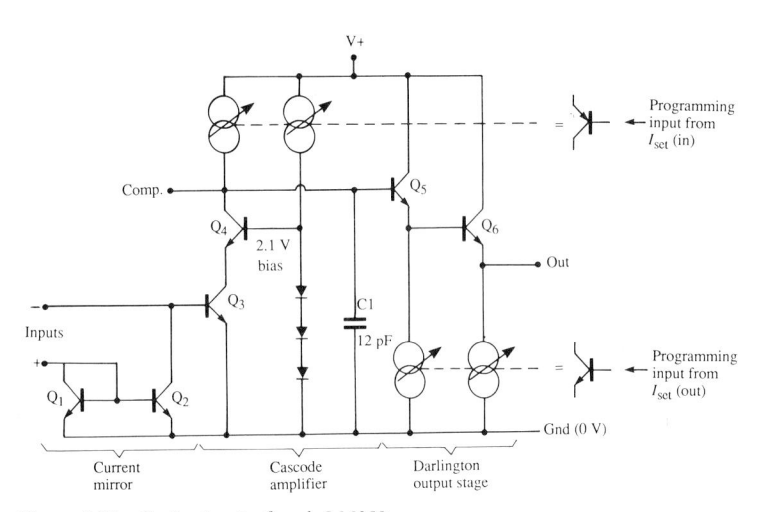

Figure 7.37   *Basic circuit of each LM359 op-amp*

Norton op-amp circuits 123

parasitic feedback effects and thus gives an excellent high-speed performance, plus a Darlington emitter follower output stage (Q_5 and Q_6).

Note that a 12 pF capacitor is internally wired between Q_4 collector (accessible at the COMP terminal) and ground, and that the Q_3–Q_4 operating current can be programmed via the $I_{set(in)}$ current of the IC's internal biasing network, thus enabling the circuit's input biasing current, slew rate, bandwidth, and supply current to be pre-set. Similarly, the operating currents of the Darlington output stage can be programmed via the $I_{set(out)}$ currents of the internal biasing network, enabling the output sink current and supply current to be pre-set.

Figure 7.38 shows the basic circuit of the IC's internal biasing network, which controls both op-amps. Thus, the $I_{set(in)}$ current can be set via a suitable resistor wired between pin-8 and the positive supply rail, and the $I_{set(out)}$ current can be set via a suitable resistor wired between pin-1 and ground-B. Alternatively, if $I_{set(in)}$ and $I_{set(out)}$ are to have equal values, the current can be set via a single resistor wired between pins 1 and 8.

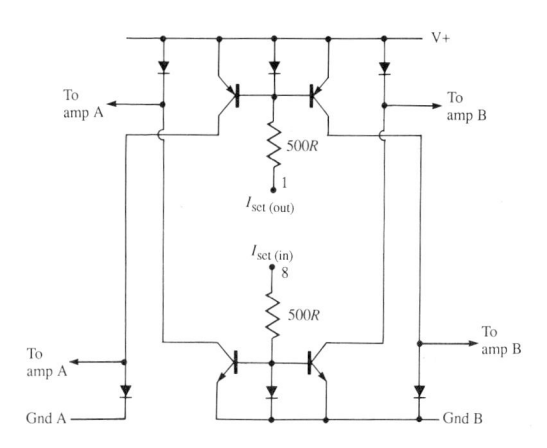

Figure 7.38   *LM359 internal programming (biasing) circuit*

Using the LM359

The LM359 is usually used in linear amplifier applications, and in such cases the design procedure involves two simple states, the first being the design of the input biasing network, and the second the selection of the programming resistor value(s).

The LM359 is biased in exactly the same way as the LM3900, using either voltage-reference biasing, supply-line biasing, or N × V_{be} biasing, as shown in *Figure 7.5* and fully described earlier in this chapter. Programming involves

124 Norton op-amp circuits

the wiring of either a single resistor between pins 1 and 8 of the IC, or of individual resistors between pin-1 and ground and between pin-8 and supply positive, as mentioned above. *Figures 7.39* to *7.41* show three typical circuits that can result from the above options, together with their relevant design formulas.

Thus, the *Figure 7.39* circuit gives inverting ac amplifier action and uses supply-line biasing, and uses individual $R_{set(in)}$ and $R_{set(out)}$ programming resistors.

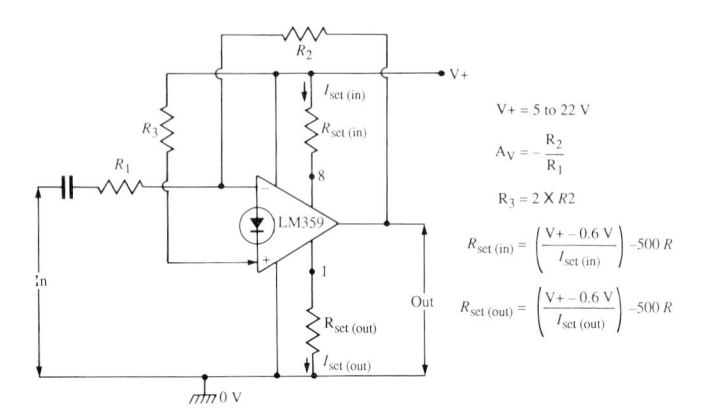

$V+ = 5 \text{ to } 22 \text{ V}$

$A_V = -\dfrac{R_2}{R_1}$

$R_3 = 2 \times R2$

$R_{set(in)} = \left(\dfrac{V+ - 0.6 \text{ V}}{I_{set(in)}}\right) - 500\,R$

$R_{set(out)} = \left(\dfrac{V+ - 0.6 \text{ V}}{I_{set(out)}}\right) - 500\,R$

Figure 7.39 *Inverting ac amplifier with supply-line biasing and individual IN and OUT programming resistors*

The *Figure 7.40* circuit also acts as an inverting ac amplifier, but uses $N \times V_{be}$ biasing and uses a single resistor for $_{in}$ and $_{out}$ programming.

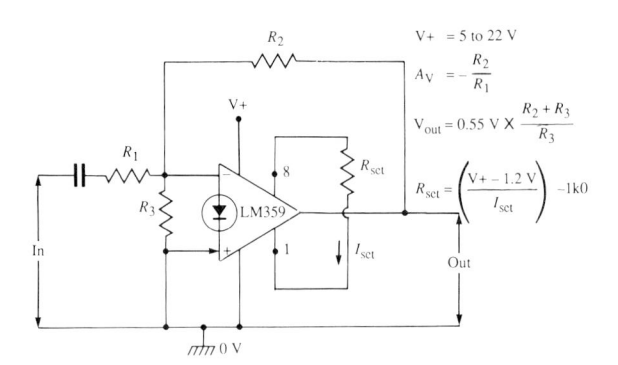

$V+ = 5 \text{ to } 22 \text{ V}$

$A_V = -\dfrac{R_2}{R_1}$

$V_{out} = 0.55 \text{ V} \times \dfrac{R_2 + R_3}{R_3}$

$R_{set} = \left(\dfrac{V+ - 1.2 \text{ V}}{I_{set}}\right) - 1k0$

Figure 7.40 *Inverting ac amplifier with $N \times V_{be}$ biasing and single resistor IN/OUT programming*

Norton op-amp circuits 125

Finally, the *Figure 7.41* circuit acts as a non-inverting ac amplifier and uses supply-line biasing and a single programming resistor.

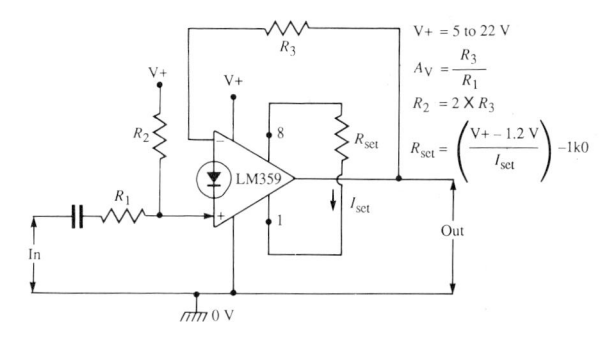

$$V+ = 5 \text{ to } 22 \text{ V}$$
$$A_V = \frac{R_3}{R_1}$$
$$R_2 = 2 \times R_3$$
$$R_{set} = \left(\frac{V+ - 1.2 \text{ V}}{I_{set}}\right) - 1\text{k}0$$

Figure 7.41 *Non-inverting ac amplifier with supply-line biasing and single resistor IN/OUT programming*

I_{set} programming

The major operating parameters of the two LM359 op-amps can be programmed via the pin-1 and pin-8 I_{set} currents of the IC. The gain-bandwidth product, the slew rate, and the inverting input bias current can be programmed via the pin-8 $I_{set(in)}$ current, and *Figures 7.42* to *7.44* show the effects of this current on the individual parameters.

The gain-bandwidth product graph of *Figure 7.42* is based on a ×100 inverting amplifier fed with a 10 MHz input signal, but is valid for all types of amplifier. Thus, with a 10 MHz input, it gives a gain of ×60 and a gain-

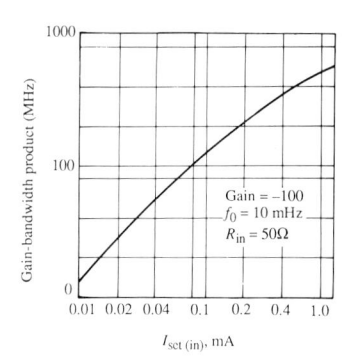

Gain = −100
f_0 = 10 mHz
R_{in} = 50Ω

$I_{set(in)}$, mA

Figure 7.42 *Gain-bandwidth product*

126 Norton op-amp circuits

bandwidth value of 600 MHz at an $I_{set(in)}$ current of 1 mA, and a gain of × 1.1 and a gain-bandwidth of 11 MHz at 0.01 mA. The gain-bandwidth of the circuit is thus directly proportional to the $I_{set(in)}$ current value.

Note that the gain-bandwidth product of the IC is also inversely proportional to the op-amp's C_{comp} value (see *Figure 7.37*), which is a fixed 12 pF but which can be increased by wiring an external capacitor between the comp terminal and ground. Thus, the gain-bandwidth values can be halved by doubling the effective C_{comp} value via an external 12 pF capacitor wired between these two points.

The slew rate (see *Figure 7.43*) of the op-amp is also directly proportional to $I_{set(in)}$ but inversely proportional to C_{comp} and can thus be varied via either of these quantities. The inverting input bias current values (*Figure 7.44*), on the other hand, are independent of C_{comp} and depend solely on the $I_{set(in)}$ values. The output sink current (*Figure 7.45*) is variable via the pin-1 $I_{set(out)}$ current and is roughly ten times that value.

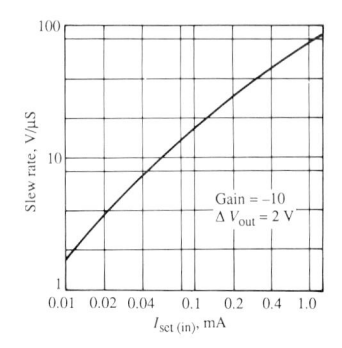

Figure 7.43 *Slew rate*

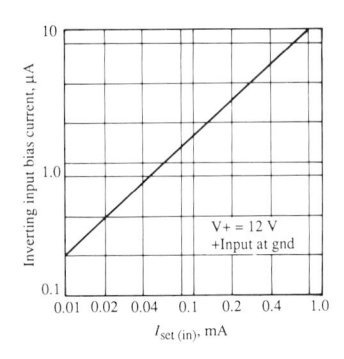

Figure 7.44 *Inverting input bias current*

Norton op-amp circuits 127

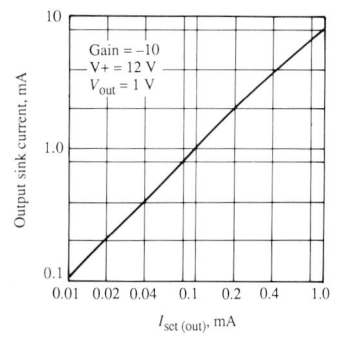

Figure 7.45 *Output sink current*

Note that, as already mentioned, the I_{set} values can either be set via individual resistors or, if both I_{set} values are equal, can be set by a single resistor wired between pins 1 and 8. If individual resistors are used, each value is determined by:

$$R_{set} = (V/I_{set}) - 500\ \Omega$$

where $V = V+ \; -0.6$ V. In this case the total current consumption of the IC (of the two op-amps) is roughly equal to:

$$I_{supply} = (27 \times I_{set(out)}) + (11 \times I_{set(in)})$$

If only a single programming resistor is used, its value is determined by:

$$R_{set} = (V/I_{set}) - 1k0$$

where $V = V+ \; -1.2$ V. In this case the total current consumption of the IC roughly equals $37 \times I_{set}$. *Figure 7.46* shows the typical consumption graph when using a 12 V supply.

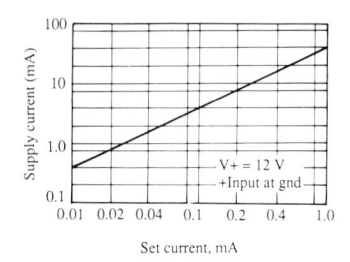

Figure 7.46 *Total supply current* ($I_{set(in)} = I_{set(out)}$)

Wideband amplifiers

The most important application of the LM359 is as a video or wideband amplifier, and *Figures 7.47* to *7.49* show three practical circuits of this type. The basic design principles of these circuits are as follows:

The *Figure 7.47* circuit is designed to be powered from a 12 V supply, and to act as a × 10 (= 20 dB) inverting amplifier that gives a bandwidth of at least 20 MHz when driven via a terminated 75 ohm line. This last requirement sets the R_1 value at 75 Ω. The input is then ac-coupled via C_1, which is shunted by C_2 to minimize its high-frequency impedance. R_2–R_3 set the circuits voltage gain. R_2 must be small, but must not significantly shunt the R_1 value; this gives R_2 a sensible compromise value of 75 Ω. To give a voltage gain of × 10, R_4 must be ten times greater than R_2, and this sets the R_4 value at 7.5 k.

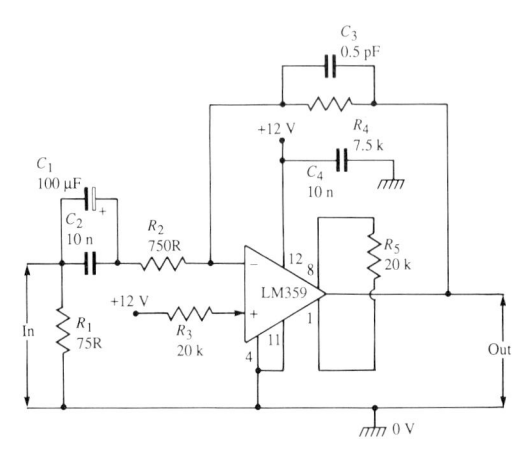

Figure 7.47 *Wideband (> 20 MHz) × 10 inverting amplifier*

To give maximal output signal swing the op-amp output must be dc biased to a quiescent value slightly below half-supply volts, and this is achieved by making R_3 a bit more than twice the R_4 value. A good compromise is 20 k, which sets the output at 5.1 V. To give the required gain and bandwidth, the op-amp needs a minimum gain-bandwidth product of 200 MHz. An I_{bias} value of 0.5 mA gives a gain-bandwidth of 400 MHz, which gives a good margin of safety, and this can be programmed by giving R_5 a value of 20 k.

To ensure a good high-frequency performance, the pin-12 supply pin is rf decoupled to ground via C_4. To give maximal bandwidth, C_3 (two twists of insulated wire) is adjusted on test. In practice this circuit gives a 3 dB bandwidth that extends from 2.5 Hz to about 30 MHz, and is absolutely flat up to 20 MHz.

Norton op-amp circuits 129

Figure 7.48 shows a non-inverting version of the wideband amplifier. In this case the gain is determined by the R_2–R_4 ratio, and the dc biasing value by the R_3–R_4 ratio. The 3 dB bandwidth of the circuit extends from 2.5 Hz to 30 MHz, and is almost flat to 20 MHz.

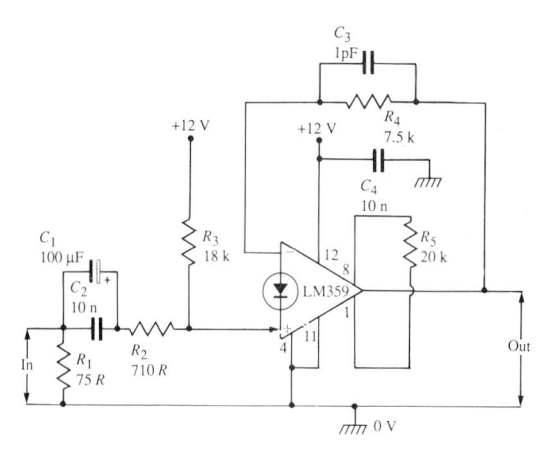

Figure 7.48 *Wideband (> 20 MHz) × 10 non-inverting amplifier*

Finally, to complete this look at Norton amplifier ICs, *Figure 7.49* shows how both op-amps of an LM359 IC can be cascaded to make a general-purpose wideband amplifier with a nominal gain of ×1000 and a 3 dB bandwidth that extends from 10 Hz to 8 MHz. In this case the op-amps are each wired in the inverting amplifier mode, with ×33 gain set by the R_3/R_1 or R_6/R_4 ratio, and use N × V_{be} biasing, with the N ratio set by R_3/R_2 or R_6/R_5.

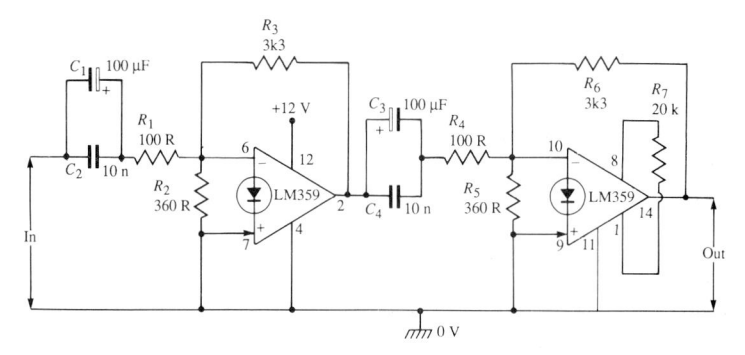

Figure 7.49 *High-gain (× 1000) general-purpose wideband (8 MHz) amplifier*

8 CA3080 OTA circuits

The CA3080 is a special type of op-amp known as an operational transconductance amplifier (OTA) and is of particular value in making voltage- or current-controlled amplifiers, or micro-power voltage comparators or oscillators, etc. OTAs have operating characteristics very different from conventional op-amps, and *Figure 8.1* illustrates the major differences between these two types of device.

Figure 8.1(a) shows the basic symbol and formulas of the conventional op-amp, which is essentially a voltage amplifying device. It has differential input terminals and gives an output of $A_o \times (e_1 - e_2)$, where A_o is the open-loop voltage gain of the op-amp and e_1 and e_2 are the signal voltages at the non-inverting and inverting input terminals respectively. Note that the open-loop voltage gain of this op-amp is fixed, and the device has a high input impedance and a low output impedance.

Figure 8.1(b) shows the basic symbol and formulas of an OTA, which is essentially a voltage-to-current amplifier. It has differential voltage input

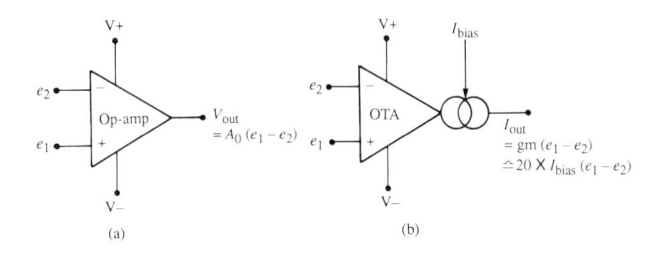

Figure 8.1 *A conventional op-amp (a) is a fixed-gain voltage-amplifying device. An OTA (b) is a variable-gain voltage-to-current amplifier*

CA3080 OTA circuits   131

terminals (like a conventional op-amp) but, as indicated by the constant-current symbol on its output, these input voltages produce a high-impedance output in the form of a current with a value of gm × $(e_1 - e_2)$, where gm is the transconductance or voltage-to-current gain of the device and can be controlled by (and is directly proportional to the value of) an external bias current fed into the I_{bias} terminal. In the CA3080, I_{bias} can be varied from 0.1 μA to 1 mA, giving a 10000:1 gain-control range.

An OTA is a very versatile device. It can, for example, be made to act like a normal op-amp by simply wiring a suitable load resistance to its output terminal (to convert its output current into voltage). Again, since the magnitude of I_{bias} can easily be controlled by an external voltage and a series resistor, the OTA can easily be used as a voltage-controlled amplifier (VCA), oscillator (VCO) or filter (VCF), etc. Note that the total current consumption of the CA3080 OTA is only twice the I_{bias} value (which can be as low as 0.1 μA), enabling the device to be used in true micro-power applications.

The best known versions of the OTA are the CA3080 and the LM13600 or LM13700. The LM13600 and LM13700 are dual second generation OTAs with built-in output-buffer stages, and are fully described in Chapter 9 of this volume. The CA3080 is a first-generation OTA, and is the exclusive subject of this chapter. *Figure 8.2(a)* shows the connections of the 8-pin DIL 'E' version of the CA3080. *Figure 8.2(b)* shows its internal circuit, and *Figure 8.3* lists its basic parameter values.

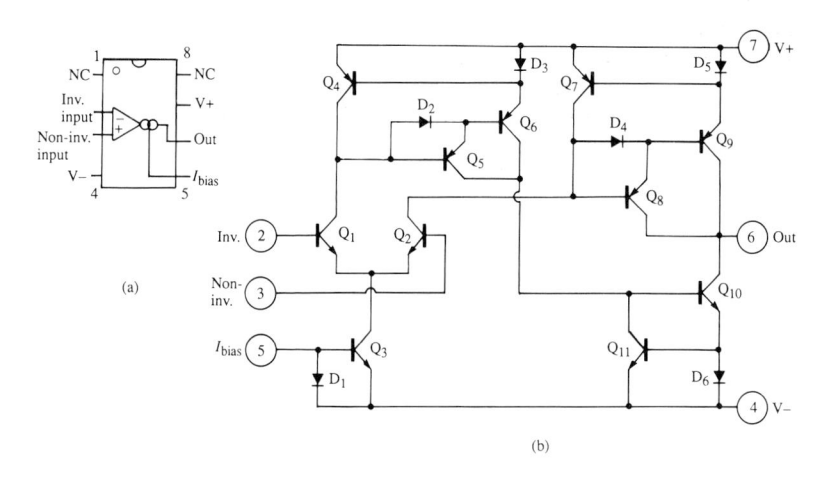

Figure 8.2 *Connections (a) of the 8-pin DIL 'E' version of the CA3080, and (b) its internal circuit*

132 CA3080 OTA circuits

Characteristic	Limits
Supply voltage range	+4 V to +30 V DC or ±2 V to ±15 V
Maximum differential input voltage	±5 V
Power dissipation	125 mW maximum
Input signal current	1 mA maximum
Amplifier bias current	2 mA maximum
Output short-circuit duration	Indefinite
Forward transconductance, gm	9600 µmho typical
Open loop bandwidth	2 MHz
Unity-gain slew rate	50 V/µS
Common-mode rejection ratio	110 dB typical

Figure 8.3 *Basic parameters/limits of the CA3080E*

CA3080 basics

The CA3080 is a fairly simple device and consists of one differential amplifier and four current mirrors. *Figure 8.3* shows the basic circuit and formulas of its differential amplifier. Its emitter current (I_c) is equal to the sum of the two collector currents (I_a and I_b); when V_{in} is zero, I_a and I_b are equal and have a value of $I_c/2$. When V_{in} has a value other than zero (up to ±25 mV maximum) the I_a and I_b currents differ and produce an $I_b - I_a$ value of $V_{in} \times $ gm, where gm is the OTa's transconductance, is directly proportional to I_c, and has a typical mho value of about $20 \times I_c$.

The *Figure 8.4* circuit is of little value on its own, and in the CA3080 it is turned to good use by using a simple current mirror to externally control its I_c value (and thus the gm of the OTA), and by using another three current mirrors to extract the difference between the I_a and I_b currents and make this difference current available to the outside world.

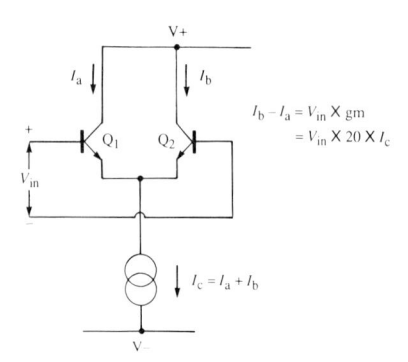

Figure 8.4 *Basic circuit and formulas of the differential amplifier of the CA3080*

CA3080 OTA circuits 133

A current mirror (CM) is a 3-terminal circuit that, when provided with an external input bias current, produces an in-phase current of identical value at its output terminals, as shown in *Figure 8.5*. Some CMs act as current sinks, as shown in *Figure 8.5(a)* and others as current sources, as in *Figure 8.5(b)*. When a CM source and a CM sink are connected as shown in *Figure 8.6* and powered from split supply rails, they generate a differential ($I_{source}-I_{sink}$) current in any external load that is connected to the 0 V rail.

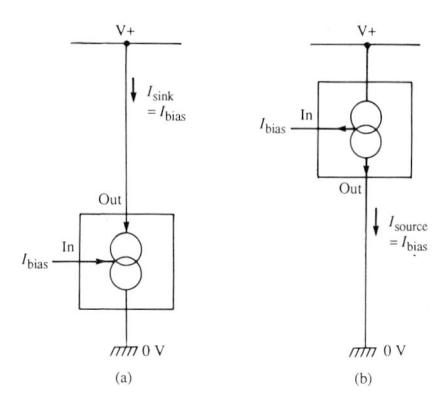

Figure 8.5 *Symbols representing (a) a current-mirror sink and (b) a current-mirror source*

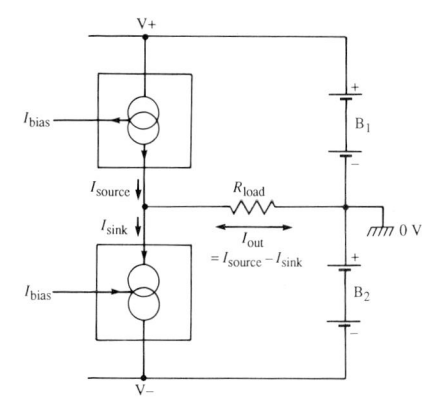

Figure 8.6 *When two current mirrors are wired as shown, they generate a differential current in an external load*

134 CA3080 OTA circuits

Figure 8.7 shows the actual circuits of two sink-type current mirrors. In the simplest of these (*Figure 8.7(a)*) a diode-connected transistor (QA) is wired across the base-emitter junction of a second, closely matched, transistor that is integrated on the same chip. The input current is fed to the bases of both matched transistors and thus divides equally between them. Suppose these transistors have current gains of × 100 and are each drawing base currents of 5 μA. In this case they each draw collector currents of 500 μA. Note, however, that the QA collector current is drawn from the circuit's input current, which thus equals 500 μA plus (2 × 5 μA), or 510 μA, and that the QB collector current is the output or mirror current of the circuit. The input and output currents of this circuit are thus almost identical (within a few per cent), irrespective of the input current magnitude.

In practice, the input/output current ratio of the above circuit depends on the close gain matching of the two transistors, and this can actually vary by several per cent. *Figure 8.7(b)* shows an improved current mirror circuit that is less sensitive to current gain variations and also gives an improved (greater) output impedance.

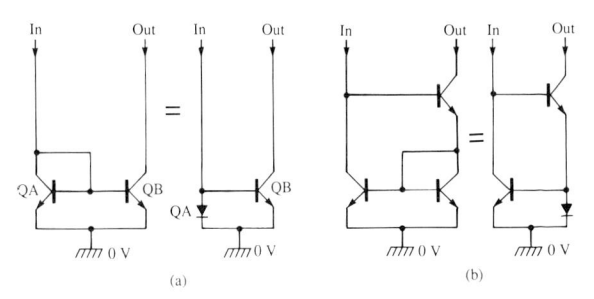

Figure 8.7 *Examples of simple sink-type current mirrors*

Figure 8.8 shows how the differential amplifier and four current mirrors are interconnected in the CA3080 to make a practical OTA. Bias current I_{bias} controls the emitter current, and thus the gm, of the Q_1–Q_2 differential amplifier via CM C. The collector currents of Q_1 and Q_2 are mirrored by CMA and CMB respectively, and then fed into the bias and sink terminals respectively of CMD, so that the externally available output current of the circuit is equal to I_b–I_a.

Looking back to *Figure 8.2(b)*, which shows the actual internal circuit of the CA3080, the reader should now have little difficulty in working out the functions of the individual circuit elements. Q_1 and Q_2 form the differential amplifier, with D_1–Q_3 making up CMC of *Figure 8.8*, and CMD comprises D_6–Q_{10}–Q_{11}. Current mirrors CMA (Q_4–Q_5–Q_6–D_2–D_3) and CMB (Q_7–Q_8–Q_9–D_4–D_5) are slightly more complex, using Darlington pairs of transistors, plus speed-up diodes, to improve their performances.

CA3080 OTA circuits   135

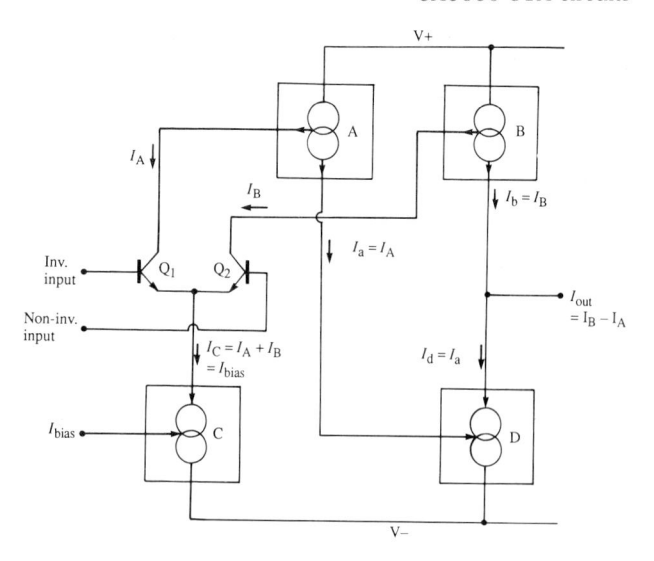

Figure 8.8 *The CA3080 comprises one differential amplifier and four current mirrors*

Some finer points

All of the major operating parameters of the CA3080 are adjustable and depend on the value of I_{bias}. The maximum (short circuit) output current is equal to I_{bias}, the total operating current of the OTA is double the I_{bias} value, and the input bias currents drawn by pins 2 and 3 typically equal $I_{bias}/200$.

The transconductance (gm) and the input and output impedance values also vary with the I_{bias} value, as shown in the graph of *Figure 8.9*, which shows

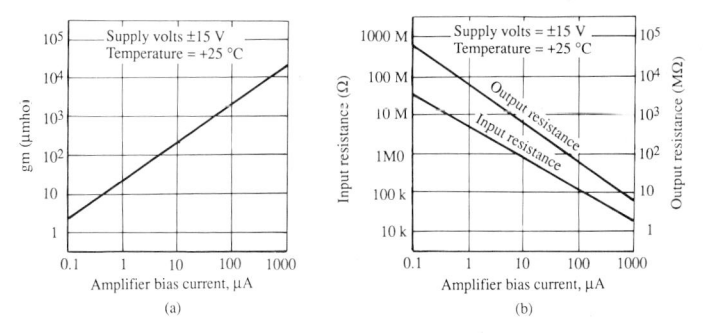

Figure 8.9 *The transconductance (a) and the input and output resistances (b) of the CA3080 vary with the bias current value*

136 CA3080 OTA circuits

typical parameter values when the IC is driven from split 15 V supplies at an ambient temperature of +25°C. Thus, at a bias current of 10 μA, gm is typically 200 μmho, and input and output impedance are 800 k and 700 MΩ respectively. At 1 mA bias the values change to 20 mmho, 15 k and 7 MΩ respectively.

The available output voltage swing of the IC depends on the values of I_{bias} and any external load resistor connected to the OTA output. If the load impedance is infinite, the output can swing to within 1V5 of the positive supply rail and to within 0V5 of the negative rail. If the impedance is finite, the peak output voltage swing is limited to $I_{bias} \times R_L$. Thus, at 10 μA bias with a 100 kΩ load, the available output voltage swing is ±1V0.

The slew rate (and bandwidth) of the IC depend on the value of I_{bias} and any external loading capacitor connected to the output. The slew rate value, in volts per microseconds, equals I_{bias}/C_L, where C_L is the loading capacitance value in pF, and the I_{bias} value is in microamperes. With no external loading capacitor connected the maximum slew rate of the CA3080 is about 50 V/μs.

Basic circuits

The CA3080 is a very easy IC to use. Its pin-5 I_{bias} terminal is internally connected to the pin-4 negative supply rail via a base-emitter junction, so the biased voltage of the terminal is about 600 mV above the pin-4 value. I_{bias} can thus be obtained by connecting pin-5 to either the common rail or the positive supply rail via a suitable current limiting resistor.

Figures 8.10 and *8.11* show two simple ways of using the CA3080 as a linear amplifier with a voltage gain of about 40 dB. The *Figure 8.10* circuit acts as a direct-coupled differential amplifier, and *Figure 8.11* is an ac-coupled inverting amplifier. Both designs operate from split 9-V supplies, so 17.4 V is

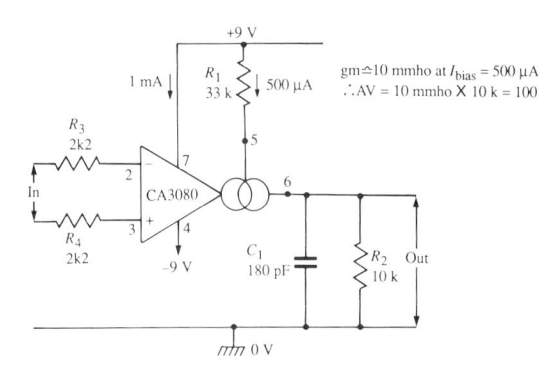

Figure 8.10 *Differential amplifier with 40 dB voltage gain*

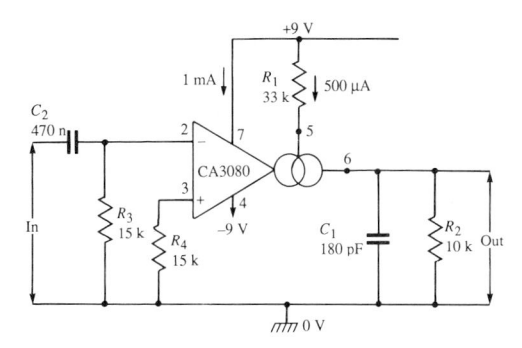

Figure 8.11   *ac-coupled 40 dB inverting amplifier*

generated across bias resistor R_1, which thus feeds an I_{bias} of about 500 μA into pin-5 and thus makes each IC consume another 1 mA from its supply rails.

At an I_{bias} value of 500 μA the gm of the CA3080 is roughly 10 mmho, so, since the outputs of the *Figure 8.10* and *8.11* circuits are loaded by a 10 k resistor (R_2), they thus give an overall voltage gain of 10 mmho × 10 k = × 100, or 40 dB. The peak current that can flow into the 10 k load is 500 μA (I_{bias}), so the peak available output voltage is ± 5V0. The output is also loaded by 180 pF capacitor C_1, giving the circuit a slew rate limit of 500 μA/180 pF, or 2.8 V/μs. The output impedance of each circuit equals the R_2 value of 10 kΩ.

Note in these two circuits that the IC is used in the open-loop mode, and that if the slew rate of the IC is not externally limited via C_1 the OTA will operate at its maximum bandwidth and slew rate. Under this condition the CA3080 may be excessively noisy, or may pick up unwanted RF signals.

In the *Figure 8.10* circuit the differential inputs are applied via series resistors R_3 and R_4, which help equalize the source impedances of the two signals and thus maintain the DC balance of the OTA. The *Figure 8.11* circuit is a simple variation of the above design, with both inputs tied to the common rail via 15 k resistors and with the input signal applied to one terminal only. With the input fed to pin-2 as shown, the circuit acts as a 40 dB inverting amplifier; alternatively, non-inverting action can be obtained by connecting the input signal to pin-3 via C_2.

Closed-loop operation

The *Figure 8.10* and *8.11* circuits are used in the open-loop mode and their voltage gains thus depend on the value of I_{bias}, which in turn depends on the value of supply rail voltage. The voltage gain of the CA3080 can be made almost independent of the I_{bias} and supply voltage values by using conventional closed-loop op-amp techniques, as shown in the micro-power 20 dB ac-coupled inverting amplifier of *Figure 8.12*.

Figure 8.12 is wired like a conventional op-amp inverting amplifier, with its voltage gain (A_v) determined primarily by the R_2/R_3 ratio (× 10, or 20 dB). This gain equation is, however, only valid when the value of an external load resistor (R_L) is infinite, since the output impedance of this design is equal to R_2/A_v, or 10 kΩ, and any external load causes this impedance to give a potential divider action that reduces the output of the circuit.

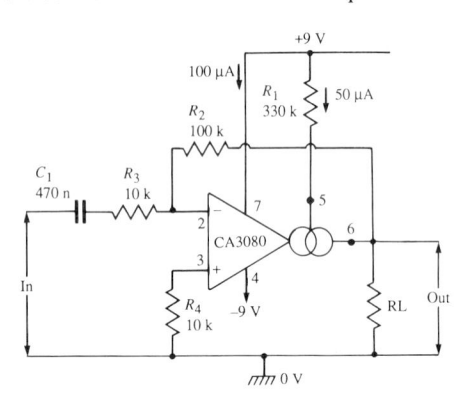

Figure 8.12 *20 dB micro-power inverting amplifier*

In *Figure 8.12*, the main function of I_{bias} is to set the total operating current of the circuit and/or the maximum available output voltage swing. With the component values shown, I_{bias} has a value of 50 μA. When R_L is infinite, the output is loaded only by R_2, which has a value of 100 k, so the maximum available output voltage swing is ±5 V 0. If R_L has a value of 10 kΩ, on the other hand, the maximum output voltage swing is ±0.5 V. This circuit can thus be designed to give any desired values of voltage gain and peak output voltage. Note that, since the IC is used in the closed loop mode, external slew rate limiting is not required.

Offset balancing

If the CA3080 is to be used as a high-gain dc amplifier, or as a wide-range variable-gain amplifier, input bias levels must be balanced to ensure that the output correctly tracks the input signals at all prevailing I_{bias} values. *Figure 8.13* shows how suitable bias can be applied to an inverting ac amplifier in which the voltage gain is variable from roughly × 5 to × 100 via RV_2, and the offset balance is pre-set via RV_1. The circuit is set up by adjusting RV_2 to its

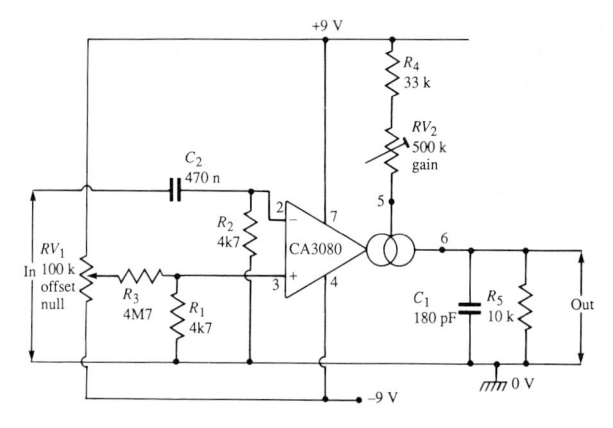

Figure 8.13 *Variable-gain (× 5 to × 100) ac amplifier*

minimum (maximum gain) value and then trimming RV_1 to give zero dc output with no ac input signal applied.

Voltage-controlled gain

The most important uses of the CA3080 are in true micro-power amplifier and oscillator applications, and in applications in which important parameters are variable via an external voltage. In the latter category, one important application is as a voltage-controlled amplifier (VCA) or amplitude modulator, in which a carrier signal is fed to an amplifier's input, and its output amplitude is controlled or modulated by another signal fed to the I_{bias} terminal. *Figure 8.14* shows a practical version of such a circuit.

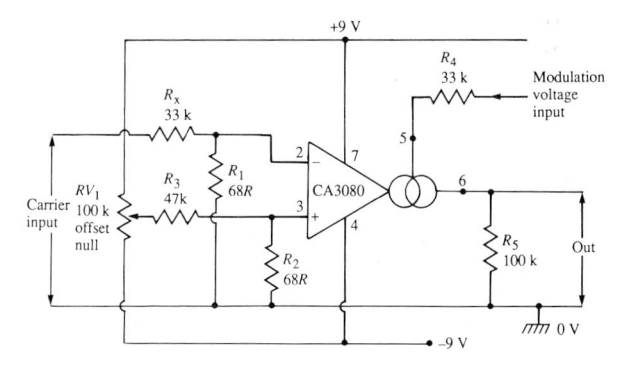

Figure 8.14 *Amplitude modulator or 2-quadrant multiplier*

140 CA3080 OTA circuits

The *Figure 8.14* circuit acts as a variable-gain inverting amplifier. Input bias resistors R_1 and R_2 have low values, to minimize the noise levels of the IC and eliminate the need for external slew-rate limiting, and offset biasing is applied to the non-inverting pin of the IC via R_3–RV_1. The carrier input signal is applied to the inverting pin of the CA3080 via potential divider R_x–R_1. When R_x has the value shown, the circuit gives roughly unity gain when the modulation input terminal is tied to the zero volts rail. The gain is ×2 when the modulation terminal is at ±9 V, and the circuit gives roughly 80 dB of signal rejection when the modulation terminal is tied to the −9 V rail.

Note that the instantaneous polarity of the output signal of the *Figure 8.14* circuit is determined entirely by the instantaneous polarity of the input signal, which has only two possible 'states'; this type of circuit is thus known as a 2-quadrant multiplier. The amplitude of the output signal is determined by the product of the input and gain-control values.

Figure 8.15 shows how the above circuit can be modified so that it acts as a ring-modulator or 4-quadrant multiplier, in which the output signal polarity depends on the polarities of both the input signal and the modulation voltage.

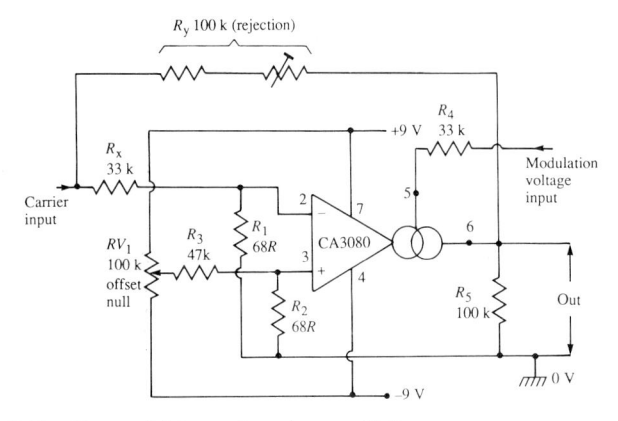

Figure 8.15   *Ring modulator or 4-quadrant multiplier*

The *Figure 8.15* circuit is identical to that of *Figure 8.14*, except that resistor network R_y is connected between the input and output terminals. The action here is such that, when the modulator input is tied to the zero volts rail, the inverted signal currents feeding into R_5 from the output of the OTA are exactly balanced by non-inverting signal currents flowing into R_5 from the input signal via R_y, so that zero output is generated across R_5. If the modulation input goes positive, the output of the OTA exceeds the current of the R_y network, and an inverted gain-controlled output is obtained. If the modulation input is negative, on the other hand, the output current of R_y

exceeds that of the OTA, and the non-inverted gain-controlled output is obtained.

Thus, both the phase and the amplitude of the output signal of this 4-quadrant multiplier are controlled by the modulation signal. The circuit can be used as a ring modulator by feeding independent AC signals to the two inputs, or as a frequency doubler by feeding identical sine waves to the two inputs.

Note that, with the R_x and R_y values shown, the *Figure 8.15* circuit gives a voltage gain of $\times 0.5$ when the modulation terminal is tied to the positive or negative supply rail; the gain doubles if the values of R_x and R_y are halved. Also note that the *Figure 8.14* and *8.15* circuits each have a high output impedance, and that in practice a voltage follower buffer stage must be interposed between the output terminal and the outside world.

Comparator circuits

The CA3080 can easily be used as a programmable or micro-power voltage comparator. *Figure 8.16* shows the basic circuit of a fast, programmable, inverting comparator, in which a reference voltage is applied to the non-inverting input terminal, and the test input is applied to the inverting terminal. The circuit action is such that the output is driven high when the test input is below V_{ref}, and is driven low when test is above V_{ref} (the circuit can be made to give a non-inverting comparator action by transposing the input connections of the IC).

With the component values shown, the I_{bias} current of the *Figure 8.16* circuit is several hundred microamperes and under this condition the CA3080 has a slew rate of about 20 V/μs and thus acts as a 'fast' comparator. When the test and V_{ref} voltages are almost identical the IC acts as a linear amplifier with a voltage gain of gm $\times$ R_2 (about $\times 200$ in this case). When the two input

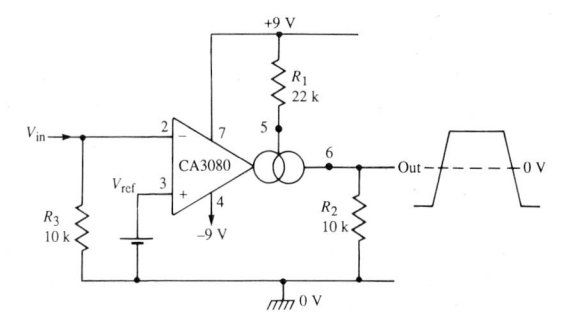

Figure 8.16 *Fast inverting voltage comparator*

142 CA3080 OTA circuits

voltages are significantly different, the output voltage limits at values determined by the I_{bias} and R_2 values. In *Figure 8.16*, the output limits at about $\pm 7V0$ when R_2 has a value of 10 k, or at ± 0.7 V when R_2 has a value of 1k0.

Figure 8.17 shows how the above circuit can be modified so that it acts as an ultra-sensitive micro-power comparator which typically consumes a quiescent current of only 50 μA but gives an output voltage that swings fully between the two supply rails and can supply drive currents of several milliamperes. Here, the CA3080 is biased at about 18 μA via R_1 but has its output fed to the near-infinite input impedance of a CMOS inverter stage made from one section of a 4007UB CMOS IC. The CA3080-plus-4007UB combination gives the circuit an overall voltage gain of about 130 dB, so input voltage shifts of only a few microvolts are sufficient to switch the output from one supply rail to the other.

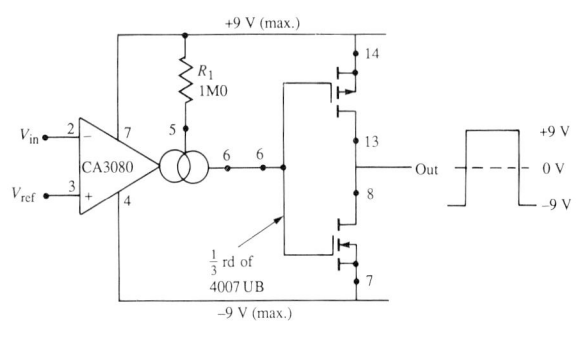

Figure 8.17 *Non-inverting micro-power voltage comparator*

Schmitt trigger circuits

The simple voltage comparator circuit of *Figure 8.16* can be made to act as a programmable Schmitt trigger by connecting the non-inverting reference terminal directly to the output of the CA3080, as shown in *Figure 8.18*. In this case, when the input is high a positive reference value of $I_{bias} \times R_2$ is generated. When V_{in} exceeds this value the output regeneratively switches low and generates a negative reference voltage of $I_{bias} \times R_2$. When V_{in} falls below this new value the output switches high again and once more generates a positive reference voltage of $I_{bias} \times R_2$. Thus, the trigger thresholds (and also the peak output voltages) of this Schmitt circuit can be precisely controlled or programmed via either I_{bias} or R_2.

CA3080 OTA circuits 143

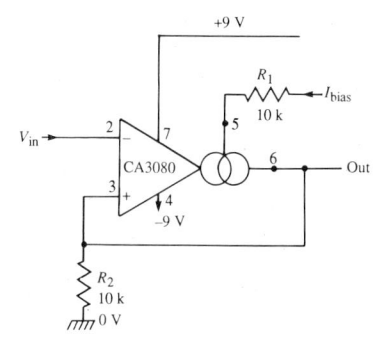

Figure 8.18 *Programmable Schmitt trigger*

Figure 8.19 shows an alternative type of Schmitt, in which the output transitions fully between the supply rail values, and the switching threshold values are determined by the R1 and R2 ratios and the values of supply voltage, V, and equal $\pm V \times R1/(R1 + R2)$.

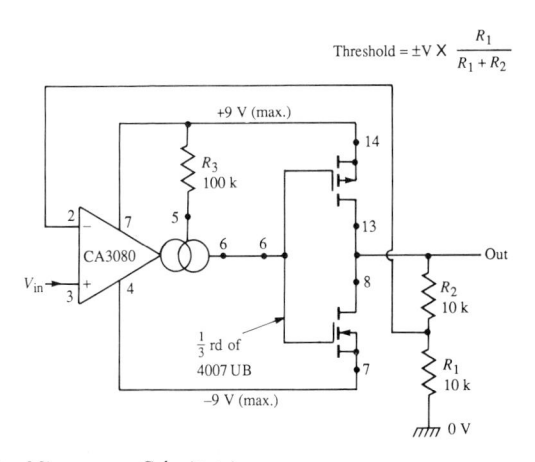

$$\text{Threshold} = \pm V \times \frac{R_1}{R_1 + R_2}$$

Figure 8.19 *Micro-power Schmitt trigger*

Astable circuits

The *Figure 8.19* Schmitt trigger circuit can be made to act as an astable multivibrator or square wave generator by connecting its output back to the non-inverting input terminal via an *R–C* time-constant network, as shown in

144 CA3080 OTA circuits

Figure 8.20. The output of this circuit switches fully between the supply rail values, is approximately symmetrical, and has a frequency that is determined by the values of R_3 and C_1, and by the ratios R_1 and R_2. The circuit action is such that, when the output is high, C_1 charges via R_3 until the C_1 voltage reaches the positive reference voltage value determined by the R_1–R_2 ratio, at which point the output switches low. C_1 then discharges via R_3 until the C_1 voltage reaches the negative reference voltage value determined by the R_1–R_2 ratios, at which point the output switches high again, and the whole process then repeats *add infinitum*.

Finally, to complete this look at the CA3080 OTA. *Figure 8.21* shows how

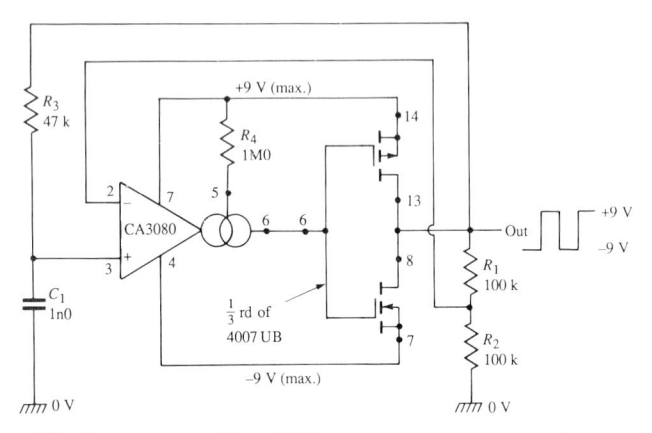

Figure 8.20 *Low-power astable multivibrator or square-wave generator*

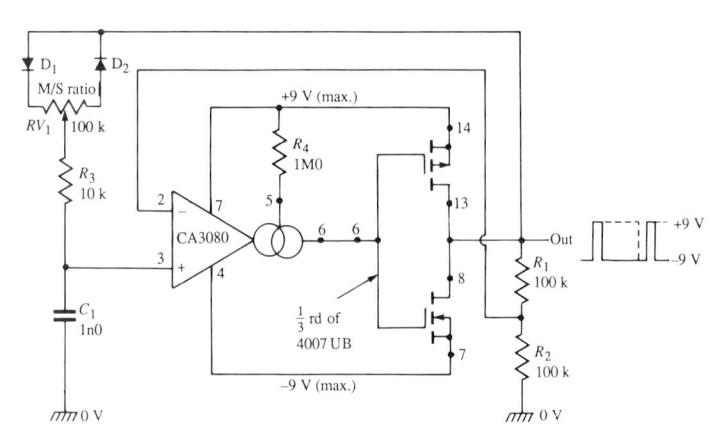

Figure 8.21 *Variable mark-space ratio astable*

the above circuit can be modified to give an output waveform with a variable mark-space ratio. In this case C_1 alternately charges via D_1–R_3 and the left half of RV_1, and discharges via D_2–R_3 and the right half of RV_1, to give a mark-space ratio that is fully variable from 10:1 to 1:10 via RV_1.

Note in the above two astable circuits that the CA3080 is biased at only a few microamperes and that the total current consumption of each design is determined primarily by the series values of R_1 and R_2 and by the value of R_3. In practice, total current consumption figures of only a few tens of microamperes can easily be obtained.

9 LM13600 OTA circuits

In Chapter 8 we explained the basic operating principles of the CA3080 operational transconductance amplifier (OTA) and showed how it can be used to make various types of voltage- or current-controlled amplifiers and micro-power voltage comparators and oscillators, etc. The CA3080 is actually a simple first generation OTA that generates fairly high signal distortion and has a high-impedance unbuffered output. In this chapter we introduce an improved second generation OTA, the LM13600 (and briefly mention its close relative, the LM13700), which does not suffer from these snags.

The LM13600 (and LM13700) is actually a dual OTA, as indicated by the pin connection diagram of *Figure 9.1*. Each of its OTAs is an improved version of the CA3080, and incorporates input linearizing diodes that greatly reduce signal distortion, and has an optional output buffer stage that can be used to give a low impedance output. The LM13600 is in fact a very versatile device, and can easily be made to act as a voltage-controlled amplifier (VCA), voltage-controlled resistor (VCR), voltage-controlled filter (VCF) or voltage-controlled oscillator (VCO), etc., as shown later in this chapter.

Linearizing diodes

The CA3080 OTA consists of a differential amplifier plus a number of current mirrors that give an output current equal to the difference between the amplifier's two collector currents, as shown in the simplified circuit of *Figure 9.2*. A weakness of this circuit is that its input signals must be limited to 25 mV peak-to-peak if excessive signal distortion is not to occur. This distortion is caused by the inherently non-linear V_{be}-to-I_c transfer characteristics of Q_1 and Q_2.

LM13600 OTA circuits 147

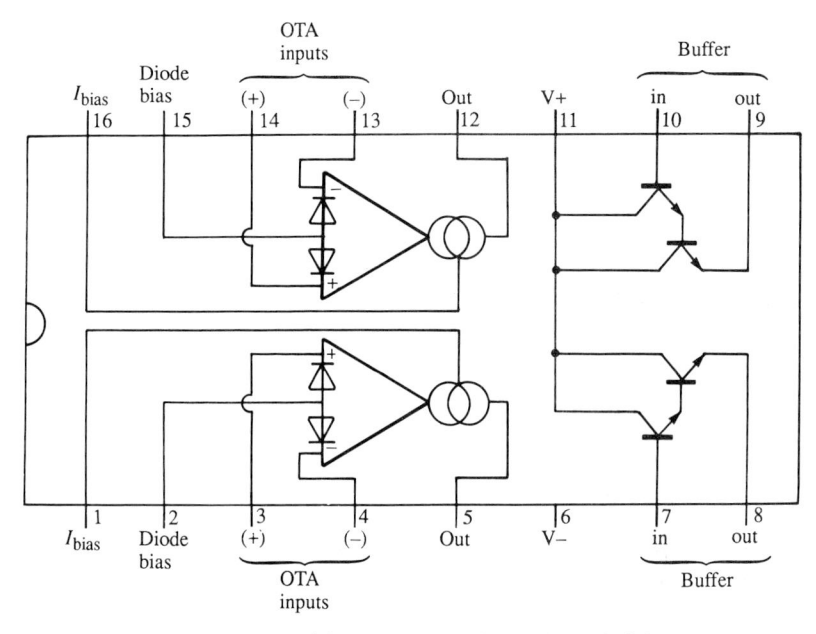

Figure 9.1 Pin connections of the LM13600 (and LM13700) dual OTA

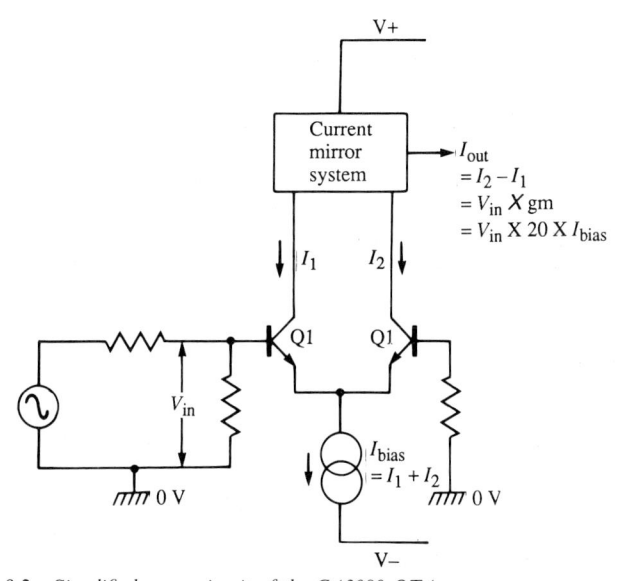

Figure 9.2 Simplified usage circuit of the CA3080 OTA

148 LM13600 OTA circuits

Figure 9.3 shows the typical transfer characteristics graph of a small-signal silicon transistor. Thus, if this transistor is biased at a quiescent collector current of 0.8 mA, an input signal of 10 mV peak-to-peak produces an output current swing of +0.2 mA to −0.16 mA and gives fairly small distortion, but an input of 30 mV peak-to-peak produces an output swing of +0.9 mA to −0.35 mA and gives massive distortion. In practice the CA3080 gives typical distortion figures of about 0.2% with a 20 mV peak-to-peak input, and a massive 8% with a 100 mV peak-to-peak input.

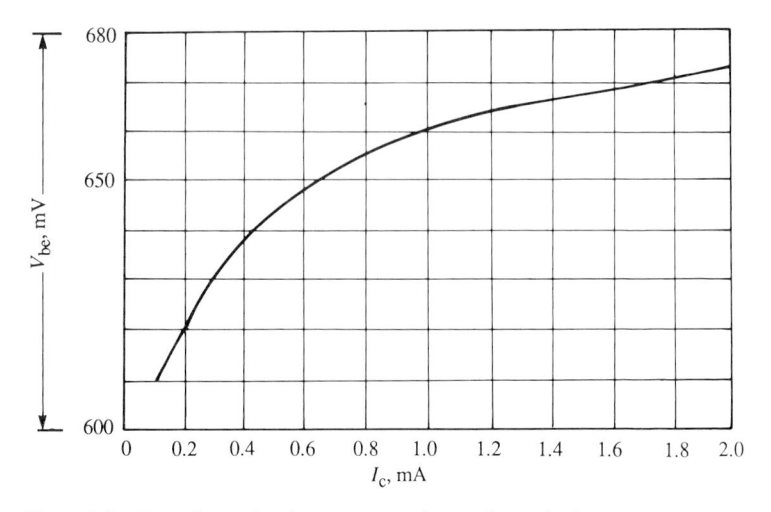

Figure 9.3 *Typical transfer characteristics of a small-signal silicon transistor*

Figure 9.4 shows the basic 'usage' circuit of one of the improved second generation OTAs of the LM13600, which is almost identical to that of the CA3080 except for the addition of linearizing diodes D_1 and D_2, which are integrated with Q_1 and Q_2 and thus have characteristics matched to those of the Q_1 and Q_2 base-emitter junctions. In use, equal low-value resistors R_1 and R_2 are wired between the inputs of the differential amplifier and the common supply line, and bias current I_D is fed to them from the positive supply rail via R_3 and D_1–D_2 and, since D_1–D_2 and R_1–R_2 are matched, divides equally between them to give R_1 and R_2 currents of $I_D/2$.

The circuit's input voltage is applied via R_4 (which is large relative to R_1) and generates input signal current I_s, which feeds into R_1 and thus generates a signal voltage across it that reduces the D_1 current to $(I_D/2) - I_s$. The I_D current is, however, constant, so the D_2 current rises to $(I_D/2) + I_s$.

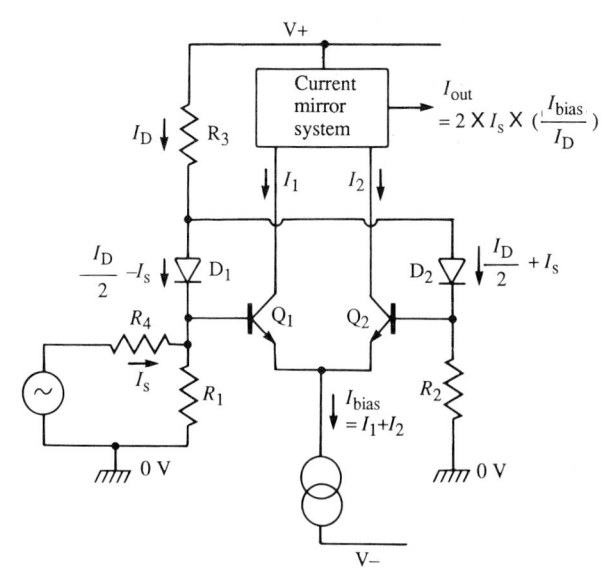

Figure 9.4 *Simplified usage circuit of an LM13600 OTA*

Consequently, the linearizing diodes of the *Figure 9.4* circuit apply heavy negative feedback to the differential amplifier and give a large reduction in signal distortion. If I_s is small relative to I_D, the output signal current of the circuit is equal to $2 \times I_s \times (I_{bias}/I_D)$; the circuit's gain can thus be controlled via either I_{bias} or I_D.

The OTAs of the LM13600 can be used as simple OTAs of the CA3080 type by ignoring the presence of the two linearizing diodes, or can be used as low-distortion amplifiers by using the diodes as shown in *Figure 9.4*. The graph of *Figure 9.5* shows typical distortion levels of the LM13600 at various peak-to-peak values of input signal voltage, with and without the use of linearizing diodes. Thus, at 30 mV input, distortion is below 0.03% with the diodes but 0.7% without them, and at 100 mV input is roughly 0.8% with the diodes but 8% without them.

Controlled impedance buffers

Figure 9.6 shows the internal circuit of each half of the LM13600 package. If this circuit is compared with that of the CA3080 shown in Chapter 8 it will be seen to be similar except for the addition of linearizing diodes D_2–D_3 and

150 LM13600 OTA circuits

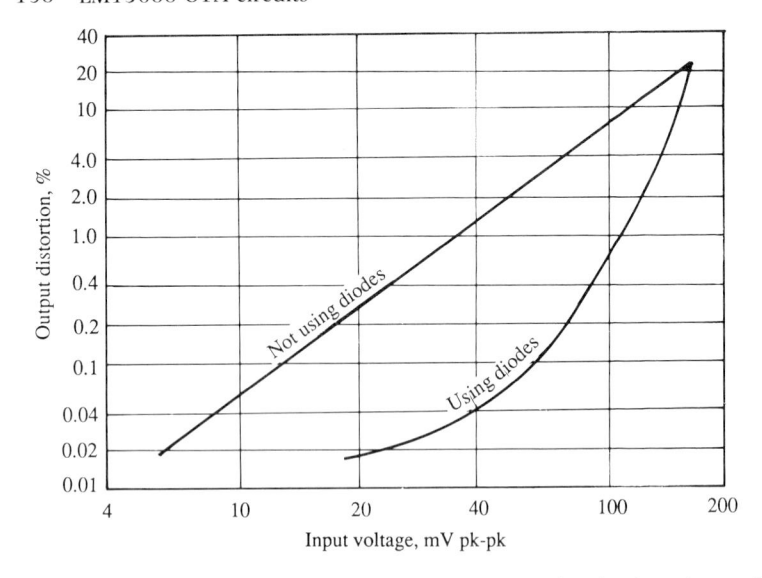

Figure 9.5 *Typical distortion levels of the LM13600 OTA, with and without the use of the linearizing diodes*

output transistors Q_{12}–Q_{13}, which are biased at a value similar to I_{bias} via Q_3. These two output transistors are connected for use as a controlled impedance Darlington emitter follower buffer stage, and can (by wiring its input to the OTA output and connecting Q_{13} emitter to the negative rail via a suitable load resistor) be used to make the high impedance output of the OTA available at a low impedance level.

Note that the output of the buffer stage is two base-emitter volt drops (about 1.2 V) below the output voltage level of the OTA, so this buffer is not suited to use in precision DC amplifier applications. The buffer is biased in parallel with I_{bias} via Q_3 to ensure that the slew-rate of the buffer increases in proportion to the I_{bias} level; the input impedance of the buffer is inversely proportional to I_{bias}.

The two OTAs of the LM13600 share common supply rails, but are otherwise fully independent. All elements are integrated on a single chip, and the OTAs thus have closely matched characteristics (gm values are typically matched within 0.3 dB), making the IC ideal for use in stereo VCA and VCF applications, etc. The standard commercial version of the LM13600 can be powered from split supply rails of up to ± 18 V, or single-ended supplies of up to 36 V. In use, I_D and I_{bias} should both be limited to 2 mA maximum, and the output currents of each buffer stage should be limited to 20 mA maximum.

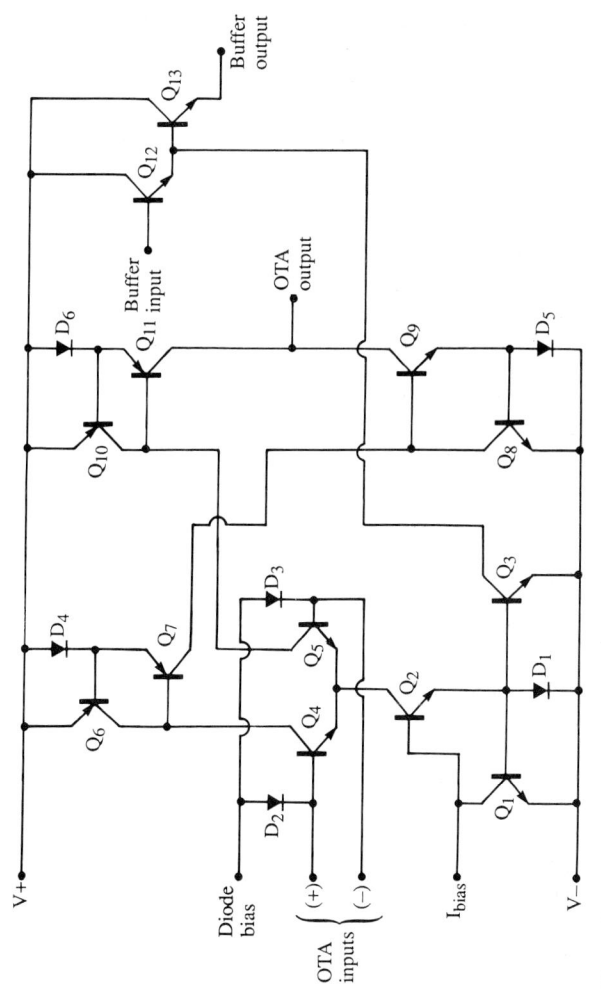

Figure 9.6 Internal circuit of one half of the LM13600 dual OTA

152 LM13600 OTA circuits

VCA circuits

Figure 9.7 shows how a practical voltage-controlled amplifier (VCA) made from one half of an LM13600 IC. Here, the input signal is fed to the non-inverting terminal of the OTA via current-limiting resistor R_4, and the high-impedance output of the OTA is loaded by R_5, which determines the peak (overload) amplitude of the output signal in the way described in Chapter 8. The output signal is made available to the outside world at a low-impedance level via the buffer stage, which is loaded via R_6.

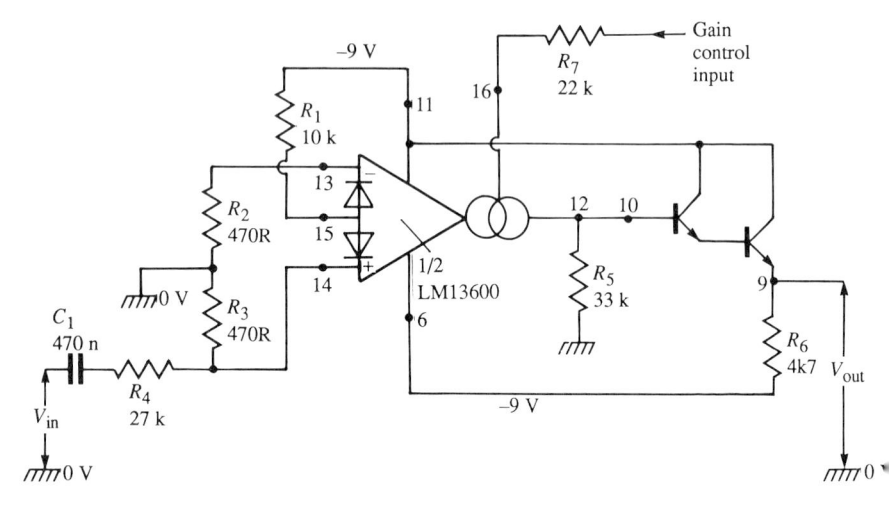

Figure 9.7 *Voltage-controlled amplifier (VCA)*

The *Figure 9.7* circuit is powered from dual 9 V supplies. The I_D current is fixed at about 0.8 mA via R_1, but I_{bias} is variable via R_7 and an external gain control voltage. When the gain-control voltage is at the negative rail value of -9 V, I_{bias} is zero and the circuit gives an overall 'gain' of -80 dB. When the gain-control is at the positive rail value of $+9$ V, I_{bias} is about 0.8 mA, and the circuit gives a voltage gain of roughly $\times 1.5$. The voltage gain is fully variable within these two limits via the gain-control input.

The two halves of the LM13600 have closely matched characteristics, making the IC ideal for use in stereo amplifier applications. *Figure 9.8* shows how two amplifiers of the *Figure 9.7* type can be used together to make a voltage-controlled stereo amplifier. Note in this case that the I_{bias} gain-control pins of the two OTAs are shorted together and fed from a single gain-control voltage and current-limiting resistor. The close matching of the OTAs

LM13600 OTA circuits   153

ensures that the gain-control currents divide equally between the two amplifiers.

Note that the *Figure 9.7* and *9.8* circuits act as non-inverting amplifiers, since their input signals are fed to the non-inverting pins of the OTAs; they can be made to act as inverting amplifiers by simply feeding the input to the inverting pins of the OTAs.

The VCA circuit of *Figure 9.7* can be used as an amplitude modulator or 2-quadrant multiplier by feeding the carrier signal to the input terminal, and the modulation signal to the gain-control input terminal. If desired, the gain-control pin can be dc biased so that a carrier output is available with no ac input signal applied. *Figure 9.9* shows a practical example of an inverting amplifier of this type. The ac modulation signal modulates the amplitude of the carries output signal.

Figure 9.10 shows how one half of an LM13600 can be used as a ring

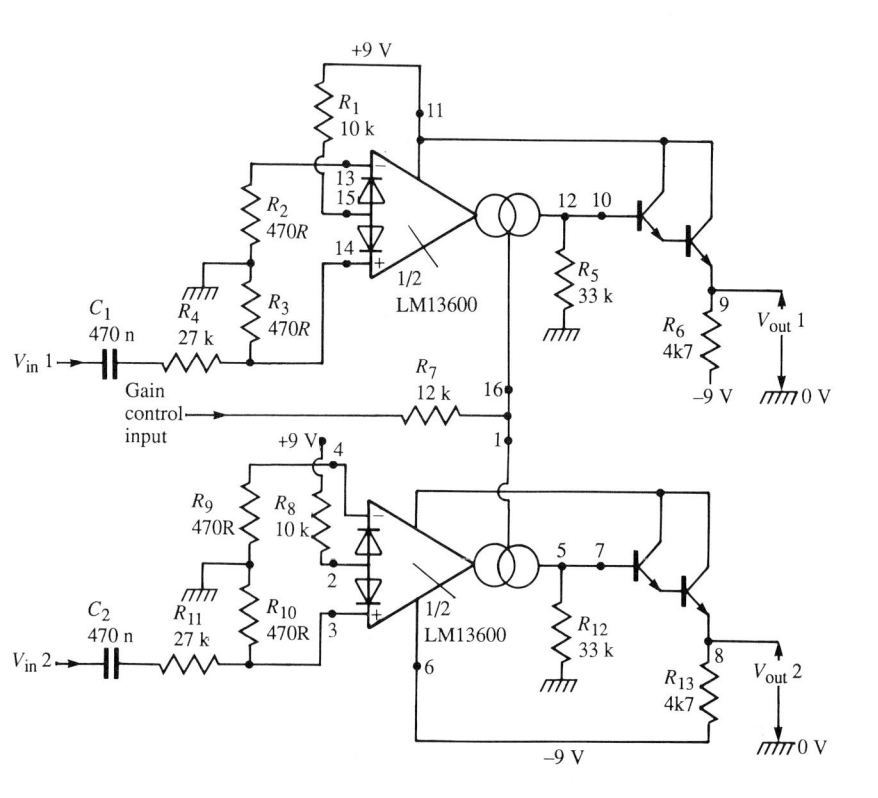

Figure 9.8 *Voltage-controlled stereo amplifier*

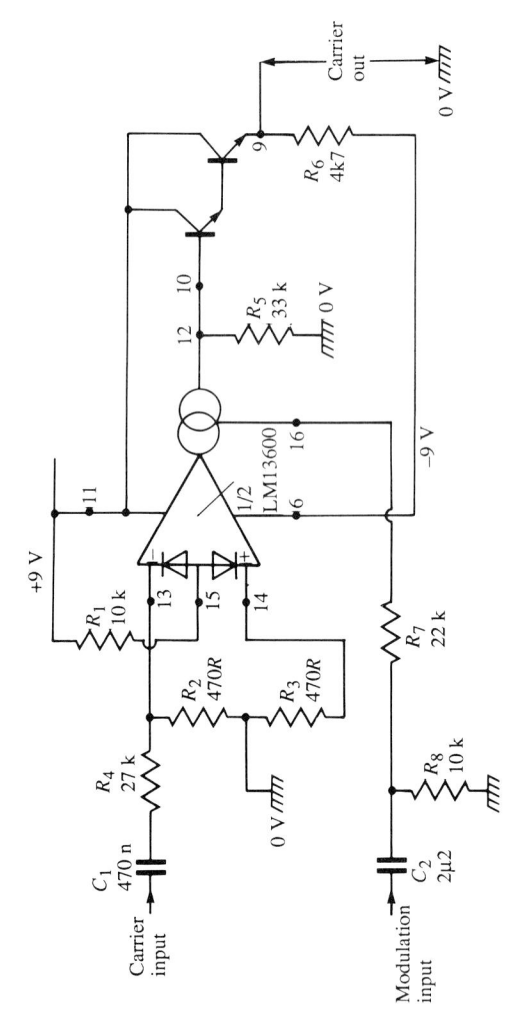

Figure 9.9 *Amplitude modulator or 2-quadrant multiplier*

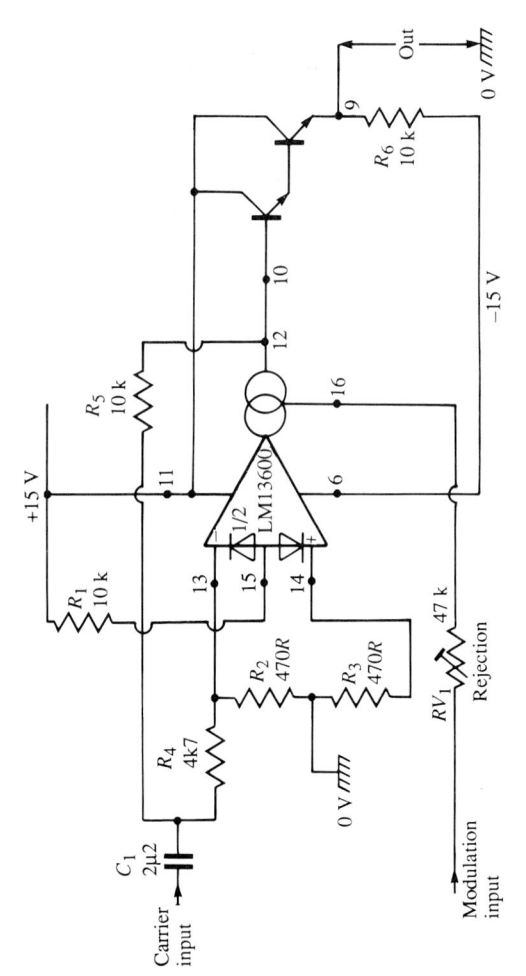

Figure 9.10 Ring modulator or 4-quadrant multiplier

156 LM13600 OTA circuits

modulator or 4-quadrant multiplier, in which zero carrier output is available when the modulation voltage is at zero (common supply rail) volts, but increases when the modulation voltage moves positive or negative relative to zero. When the modulation voltage is positive, the carrier output signal is inverted relative to the carrier input, and when the modulation voltage is negative the carrier output is non-inverted.

The *Figure 9.10* circuit is shown with values suitable for operation from dual 15 V supplies, but is essentially similar to the *Figure 9.9* circuit except that R_5 is connected between the input signal and the output of the OTA, and I_{bias} is presettable via RV_1. The basic circuit action is such that the OTA feeds an inverted (relative to the input) signal current into the bottom of R_5, and at the same time the input signal feeds directly into the top of R_5; RV_1 is pre-set so that, when the modulation input is tied to the zero volts common line, the overall gain of the OTA is such that its output current exactly balances (cancels) the direct-input current of R_5, and under this condition the circuit gives zero carrier output.

Consequently, when the modulation input goes positive, the OTA gain increases and its output signal exceeds that caused by the direct input into R_5, so an inverted output carrier signal is generated. Conversely, when the modulation input goes negative, the OTA gain decreases and the direct signal of R_5 exceeds the output of the OTA, and a non-inverted output signal is generated.

Offset biasing

The *Figures 9.7* to *9.10* circuits are shown with OTA input biasing applied via 470 R resistors wired between the two input terminals and the zero volts rail. In practice, this simple arrangement may cause the DC output level to shift slightly when the I_{bias} gain-control is varied from minimum value. If desired, this shifting can be eliminated by fitting the circuits with a presettable offset adjust control as shown in *Figure 9.11*, enabling the biasing resistance values to be varied slightly. To adjust the offset biasing, reduce I_{bias} to zero, note the dc level of the OTA output, then increase I_{bias} to maximum and adjust RV_1 to give the same dc output level.

An automatic gain control amplifier

In the *Figure 9.7* to *9.10* circuits the amplifier gain is varied by altering the I_{bias} value. A feature of the LM13600, however, is that its gain can be varied by altering either the I_{bias} or the I_D current, and *Figure 9.12* shows how I_D variation can be used to make an automatic gain control (AGC) amplifier in

LM13600 OTA circuits 157

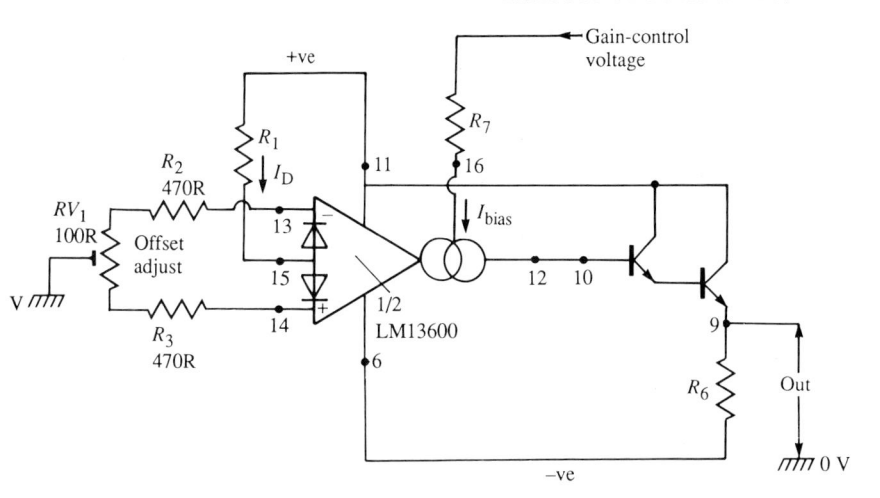

Figure 9.11 *Method of applying offset biasing to the LM13600*

which a 100:1 change in input signal amplitude causes only a 5:1 change in output amplitude.

In this circuit, I_{bias} is fixed by R_4, and the output signal is taken directly from the OTA via R_5. The output buffer is used as a signal rectifier, fed from the output of the OTA, and the rectified output is smoothed via R_5–C_2 and used to apply the I_D current to the OTA's linearizing diodes. Note, however, that no significant I_D current is generated until the OTA output reaches a high enough amplitude (3 × V_{be}, or about 1.8 V peak) to turn on the Darlington buffer and the linearizing diodes, and that an increase in I_D reduces the OTA gain and, by negative feedback action, tends to hold V_{out} at that level.

The basic zero I_D gain of this amplifier is × 40. Thus, with an input of 30 mV peak-to-peak, the OTA output of 1.2 V peak-to-peak is not enough to generate an I_D current, so the OTA operates at full gain. At 300 mV input, however, the OTA output is enough to generate significant I_D current, and the circuit's negative feedback automatically reduces the output level to 3V6 peak-to-peak, giving an overall gain of × 11.7. With an input of 3 V, the gain falls to × 2, giving an output of 6 V peak-to-peak. The circuit thus gives 20:1 signal compression over this range.

Voltage-controlled resistors

An unusual application of the LM13600 is as a voltage-controlled resistor (VCR), using the basic circuit shown in *Figure 9.13*. The basic theory here is

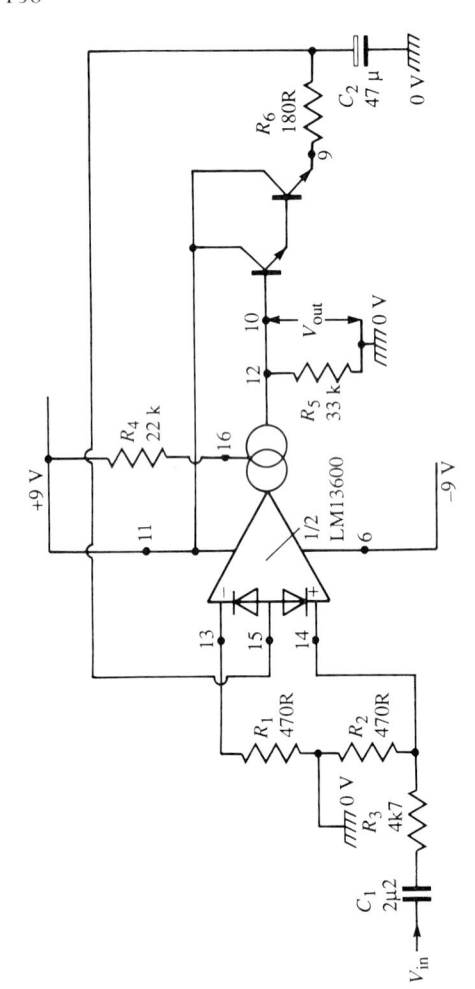

V_{in}, pk-pk	3V0	300 mV	30 mV
V_{out}, pk-pk	6V0	3V6	1V2
A_V	2	11.7	40

Figure 9.12 *Circuit and performance table of an AGC amplifier*

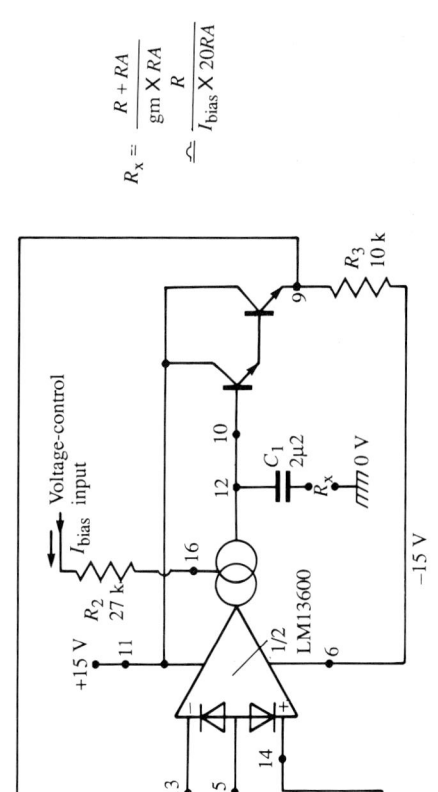

$$R_x = \frac{R + RA}{\text{gm} \times RA} \over R}$$
$$\simeq \frac{R}{I_{\text{bias}} \times 20RA}$$

Figure 9.13 *Voltage-controlled resistor, variable from 10 kΩ to 10 MΩ via I_{bias}*

quite simple. If an ac signal is applied to the R_x terminals it will feed to the OTA's inverting terminal via C_1 and the buffer stage and the R/RA attenuator, and the OTA will then generate an output current proportional to the V_{in} and I_{bias} values. Thus, since $R = E/I$, the circuit's R_x terminal acts like an ac resistor with a value determined by I_{bias}.

The effective resistance value of the R_x terminals actually equals $(R + RA)/(gm \times RA)$, where gm is roughly $20 \times I_{bias}$. This formula approximates to $R_x = R/(I_{bias} \times 20\ RA)$, so, using the component values shown in the diagram, R_x equals roughly 10 MΩ at an I_{bias} value of 1 μA, and 10 kΩ at an I_{bias} of 1 mA.

Figure 9.14 shows a similar version of the VCR, where the linearizing diodes are used to effectively improve the noise performance of the resistor, and *Figure 9.15* shows how a pair of these circuits can be used to make a floating VCR in which the input voltage is direct-coupled and may be at any value within the output voltage range of the LM13600.

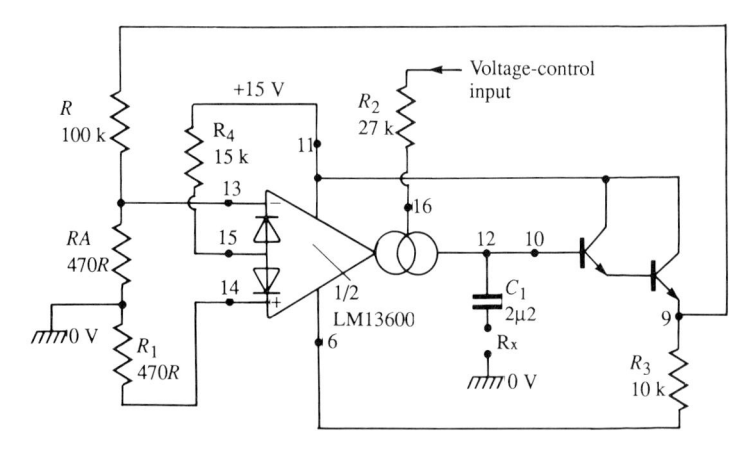

Figure 9.14 *Voltage-controlled resistor with linearizing diodes*

Voltage-controlled filters

An OTA acts basically as a voltage-controlled ac current source, in which an ac voltage is applied to the amplifier's input, and the magnitude of the output current depends on the value of this voltage and of I_{bias}. This fact can be used to implement a voltage-controlled low-pass filter by using one half of an LM13600 in the configuration shown in *Figure 9.16*, in which the values of R, C, and I_{bias} control the cut-off frequency, f_o, of the filter. The operating theory of this circuit is as follows.

161

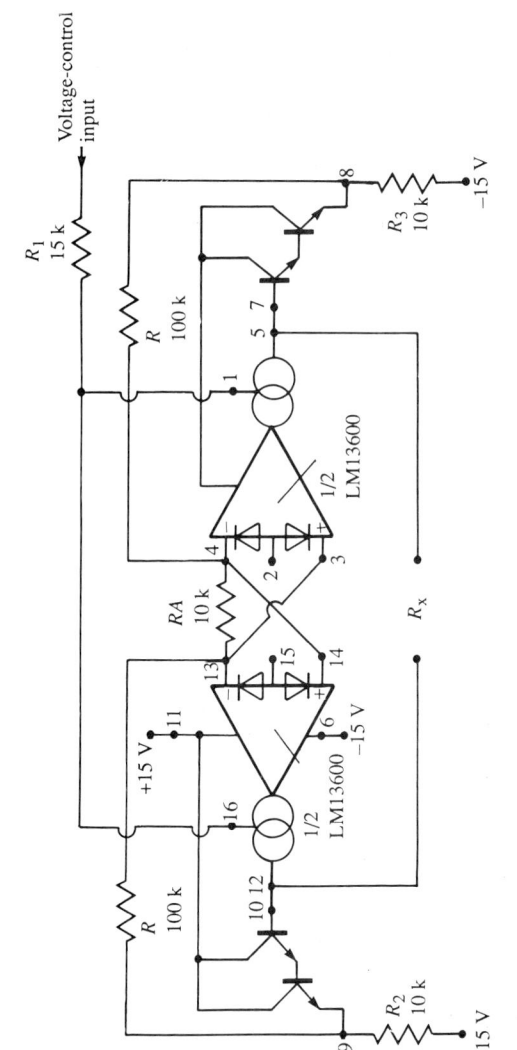

Figure 9.15 *Floating voltage-controlled resistor*

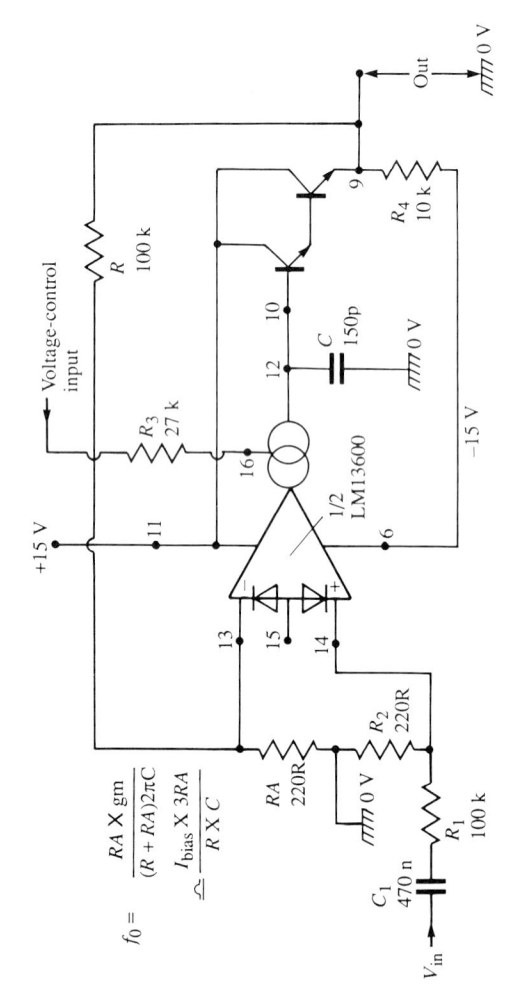

Figure 9.16 *Voltage-controlled low-pass filter covering 45 Hz to 45 kHz*

$$f_0 = \frac{RA \times gm}{(R+RA)2\pi C}$$

$$\simeq \frac{I_{bias} \times 3RA}{R \times C}$$

LM13600 OTA circuits   163

Assume initially that capacitor *C* is removed from the circuit. The input signal is applied to the OTA's non-inverting terminal via potential divider $R - R_2$, and the OTA's output is followed by the buffer stage and fed back to the inverting input via an identical divider made up of *R* and *RA*. The basic OTA thus acts as a non-inverting amplifier with a gain of R/RA but, since the input signal is fed to the OTA via a potential divider with a value equal to R/RA, the circuit acts overall as a unity-gain voltage follower.

Assume now that capacitor *C* is fitted in place. At low frequencies, *C* has a very high impedance and the OTA output current is able to fully charge it, and the circuit thus acts as a voltage follower in the way already described. As the frequency increases, however, the impedance of *C* decreases and the OTA output current is no longer able to fully charge *C*, and the output signal thus starts to attenuate at a rate of 6 dB per octave. The cut-off point of the circuit, at which the output falls by 3 dB, occurs when $XC/20I_{bias}$ equals R/RA, as implied by the formula in the diagram. With the component values shown, cut-off occurs at about 45 Hz at an I_{bias} value of 1 μA, and at 45 kHz at an I_{bias} value of 1 mA.

A similar principle to the above can be used to make a voltage-controlled high-pass filter, as shown in *Figure 9.17*. This particular circuit has, with the values shown, cut-off frequencies of 6 Hz and 6 kHz at I_{bias} currents of 1 μA and 1 mA respectively.

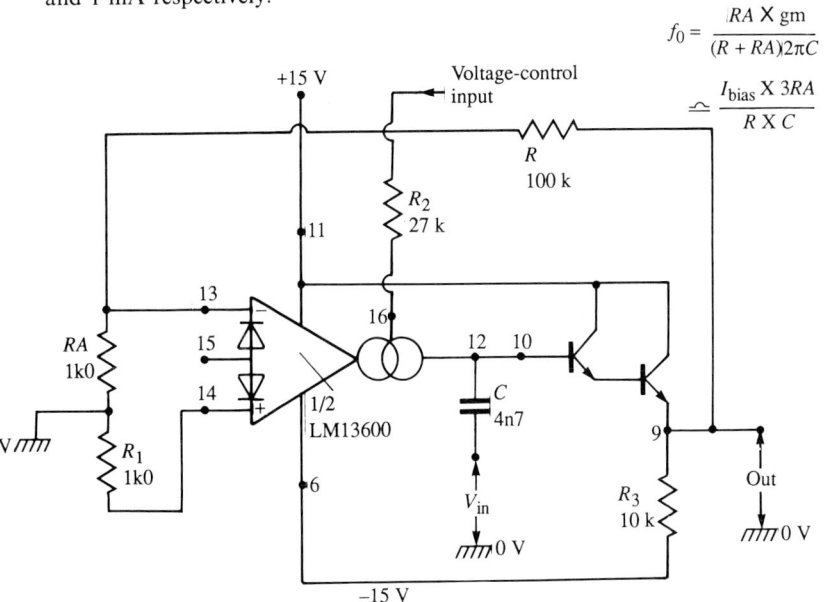

$$f_0 = \frac{RA \times g_m}{(R + RA)2\pi C}$$

$$\simeq \frac{I_{bias} \times 3RA}{R \times C}$$

Figure 9.17 *Voltage-controlled high-pass filter covering 6 Hz to 6 kHz*

Numbers of filter stages can easily be cascaded to make multi-pole voltage-controlled filters, the excellent tracking of the two sections of the LM13600 making it possible to voltage-control these filters over several decades of frequency. *Figure 9.18* shows the practical circuit of a 2-pole (12 dB per octave) Butterworth low-pass filter having cut-off frequencies of 60 Hz and 60 kHz at I_{bias} currents of 1 μA and 1 mA respectively.

Voltage controlled oscillators

To conclude this, look at applications of the LM13600, *Figures 9.19* and *9.20* show two ways of using the IC as a voltage-controlled oscillator (VCO). The *Figure 9.19* circuit uses both halves of the LM13600, and simultaneously generates both triangle and square waves. The *Figure 9.20* design uses only half of the IC, and generates square waves only.

To understand the operating theory of the *Figure 9.19* circuit, assume initially that capacitor C is negatively charged and that the square wave output signal has just switched high. Under this condition a positive voltage is developed across RA and is fed to the non-inverting terminals of the two amplifiers, which are both wired in the voltage comparator modes. This voltage makes amplifier A_1 generate a positive output current equal to bias current. I_c, and this current flows into capacitor C, which generates a positive-going linear ramp voltage that is fed to the inverting terminal of A_2 via the Darlington buffer stage, until eventually this voltage equals that on the non-inverting terminal, at which point the output of A_2 starts to swing negative. This initiates a regenerative switching action in which the square wave output terminal switches abruptly negative.

Under this new condition a negative voltage is generated across resistor RA, causing amplifier A_1 to generate a negative output current equal to I_c, and this current causes capacitor C to discharge linearly until eventually its voltage falls to a value equal to that of RA, at which point the square wave output switches high again. This process repeats *ad infinitum*, causing a triangle waveform to be generated on R_2 and a square wave output to be generated on R_4. The waveform frequency is variable via the voltage-control input, which controls the value of I_c. With the component values shown, the circuit generates a frequency of about 200 Hz at an I_c current of 1 μA, and 200 kHz at a current of 1 mA.

Finally, *Figure 9.20* shows a single-amplifier VCO circuit which generates a square wave output only. The circuit operates in a similar manner to that described above, except that C charges via D_1 and discharges via D_2, which thus generates a 'polarity' signal on the non-inverting terminal of the amplifier.

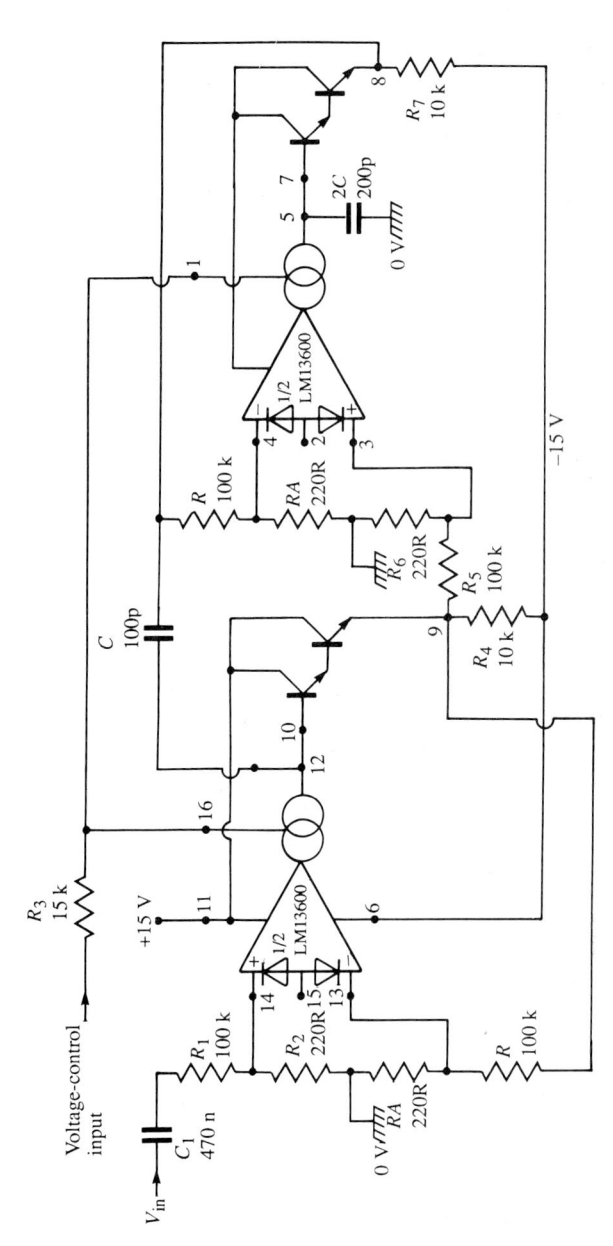

Figure 9.18 *Voltage-controlled 2-pole Butterworth low-pass filter covering 60 Hz to 60 kHz*

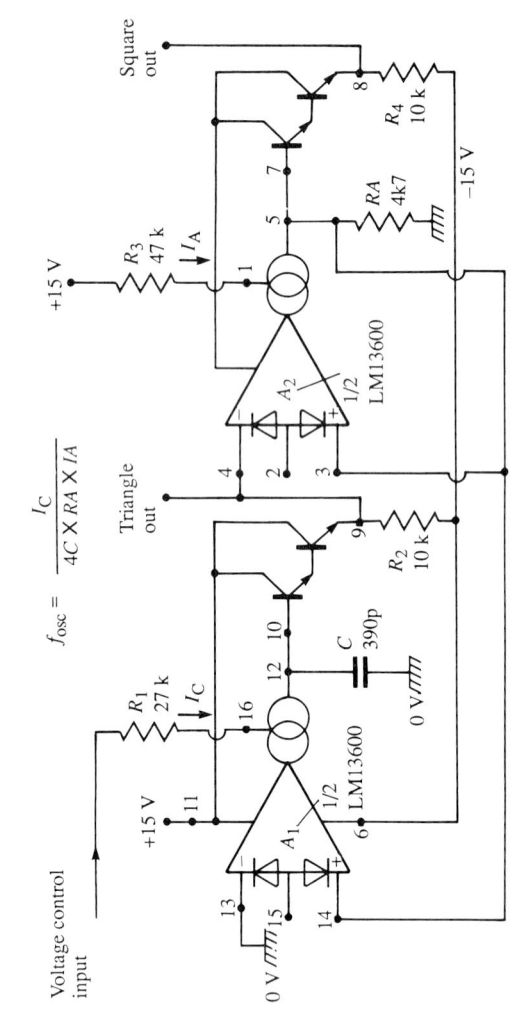

Figure 9.19 Triangle/square wave VCO covering 200 Hz to 200 kHz

$$f_{osc} = \frac{I_C}{4C \times RA \times IA}$$

LM13600 OTA circuits   167

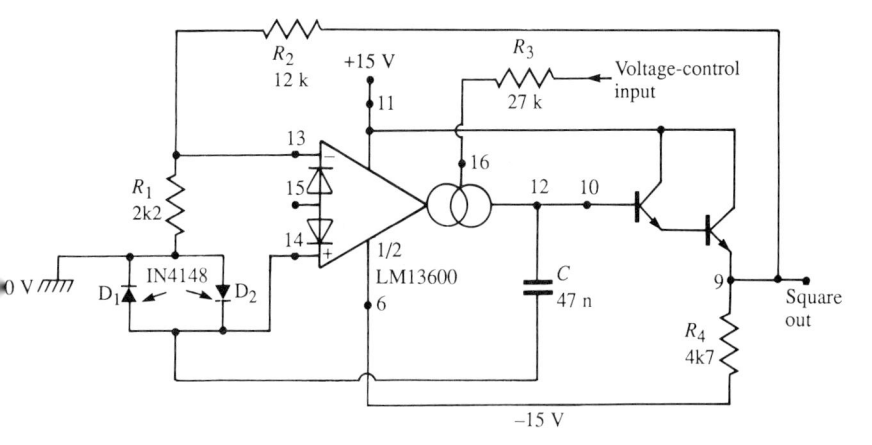

Figure 9.20 *Single-amplifier VCO*

The LM13700

All the circuits in this chapter are shown designed around the LM13600 dual OTA IC. The LM13700 IC is an almost identical device that is sometimes more readily available, and it can be used as a pin-for-pin replacement for the LM13600 in all of these circuits; it differs from the LM13600 only in details of its output buffer stages.

The internal circuitry of the LM13700 is in fact identical to that shown in *Figure 9.6* except for the elimination of Q_3, which gives the controlled impedance action to the Darlington emitter follower buffer stage in the LM13600. The output buffer stage of the LM13700 thus acts as a simple high-impedance Darlington emitter follower circuit.

10 LM10 Op-amp and reference circuits

This chapter deals with a unique bipolar linear IC known as the LM10, which houses a high-performance op-amp and a precision voltage reference in an 8-pin TO5 package, and which can operate from supplies as low as 1.1 V or as high as 45 V, and draws a typical quiescent current of only 270 μA. Its op-amp output can swing to within a few millivolts of either supply rail, or can be used to deliver output currents as high as 20 mA.

The LM10 can be used in a wide variety of both conventional and unique op-amp applications, but is particularly suited to applications calling for the use of single-ended or very-low-voltage power supplies or very low power consumption: it can also be used in many voltage-reference and power-supply applications.

LM10 basics

The LM10 contains a high-performance op-amp, plus a precision 200 mV band-gap voltage reference and a 2-input buffer amplifier that has one input terminal tied to the reference unit. *Figure 10.1* shows the outline and pin notations of the LM10, and *Figure 10.2* shows its functional diagram.

The op-amp section has a pnp differential input stage that can accept input signals down to ground volts, and has a complementary class-B output stage that can swing to within 50 mV of either supply rail at 50 μA load current, or within 400 mV at 20 mA load current. Its input is well protected (via internal current-limiting resistors) against damage from excessive voltage, and its output is fully protected against thermal overload and short-circuit damage. It can be used with either dual or single-ended power supplies, in the same way as any normal op-amp.

LM10 op-amp/reference circuits   169

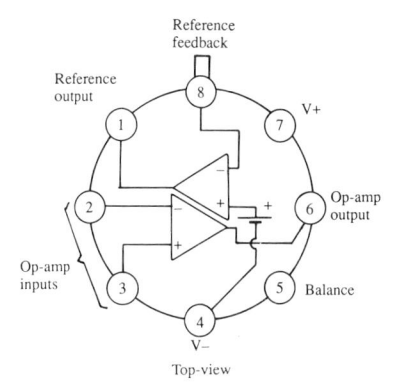

Figure 10.1   TO5 outline and pin notations of the LM10

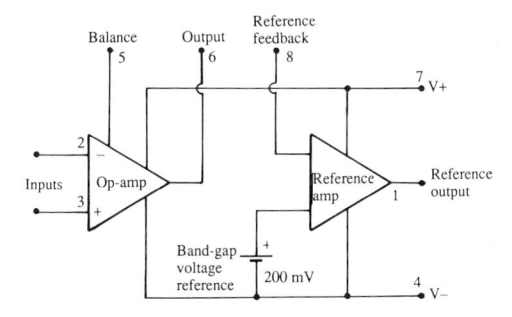

Figure 10.2   Functional diagram of the LM10

The 200 mV voltage reference section of the LM10 is a precision band-gap type with a basic accuracy better than 5% and with a temperature coefficient better than 0.003%/°C. The reference is externally accessible only via the output of the 2-input buffer amplifier, which can have its gain varied from unity upwards via suitable feedback networks, giving actual output references in the range 200 mV to 40 V.

There are five individual members of the LM10 family of devices, and these are categorized by their operating temperature ranges (LM10, LM10B, or LM10C) and their maximum supply voltage ranges of either 7 V (L suffix) or 45 V. The LM10C is a relaxed-specification commercial version of the 45 V unit, and is recommended for use in all LM10 applications shown in this chapter. *Figure 10.3* lists the main parameters of the LM10 family of devices. Note that all parameters except the unity gain bandwidth (0.3 MHz) and the slew rate (0.15 V/µs) are exceptionally good (the op-amp is not designed for high-frequency operation).

170 LM10 op-amp/reference circuits

Parameter	LM10	LM10B	LM10C	LM10BL	LM10CL	Units
Operating temperature range	−55 to 135	−25 to 85	0 to 70	−25 to 85	0 to 70	°C
Maximum total supply volts	45	45	45	7	7	V
Total quiescent current (typical)	270	270	300	260	280	µA
Typical op-amp performance at 25 °C						
Input offset voltage	0.3	0.3	0.5	0.3	0.5	mV
Input offset current	0.25	0.35	0.4	0.1	0.2	nA
Input bias current	10	10	12	10	10	nA
Common-mode rejection	102	102	102	102	102	dB
Supply-voltage rejection	96	96	96	96	96	dB
Unloaded voltage gain	400	400	400	300	300	V/mV
Loaded voltage gain ($R_L = 1k\Omega$)	130	130	130	30	30	V/mV
Unity gain bandwidth	0.3	0.3	0.3	0.3	0.3	MHz
Slew rate	0.15	0.15	0.15	0.15	0.15	V/µS
Typical performance of reference at 25 °C						
Bandgap voltage reference	200	200	200	200	200	mV
Reference voltage accuracy	2.5	2.5	5	2.5	5	±%
Line regulation	0.001	0.001	0.001	0.001	0.001	%/V
Load regulation (0 to 1mA)	0.01	0.01	0.01	0.01	0.01	%
Thermal drift	0.002	0.002	0.003	0.002	0.003	%/°C
Feedback bias current	20	20	22	20	22	nA
Amplifier gain	75	75	70	70	70	V/mV

Figure 10.3 *Parameter values of the LM10 series of devices*

Power supply connections

The LM10 is very easy to use. It can be powered from either grounded or fully-floating single-ended or dual supplies, and can use total voltages anywhere in the range 1.1 V to 45 V. *Figures 10.4 to 10.8* show a few ways of powering it in basic op-amp applications.

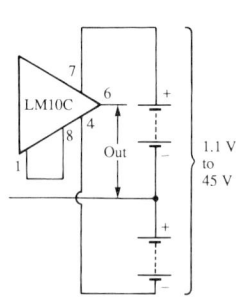

Figure 10.4 *Method of powering the LM10 for conventional split-supply operation*

Figures 10.4 and *10.5* show ways of powering it from dual supplies, for op-amp applications in which the inputs are referenced to the zero volts rail and the output can swing between the positive and negative supply rail voltages. The *Figure 10.4* circuit uses two independent supply rails, and the *Figure 10.5* design uses two rails derived from a single source.

LM10 op-amp/reference circuits 171

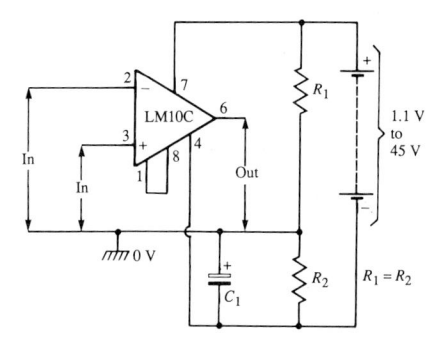

Figure 10.5 *Method of powering the LM10 for split-supply operation, using a single supply source*

Figure 10.6 shows the standard way of powering the LM10 from a single pair of supply rails. A useful feature of this circuit is that the op-amp can handle input signals right down to ground volts.

Shunt operation

Figures 10.7 and *10.8* show two unique and very useful ways of powering the LM10 from a single pair of supply rails. In each of these configurations the op-amp output terminal is shorted directly to the IC's positive supply pin, so that the output shunts the devices supply current, and a current-limiting resistor is wired in series with one of the IC's supply leads.

The LM10 op-amp has an output drive current capacity two orders of magnitude greater than the IC's normal quiescent current. This factor, combined with the op-amp's excellent supply-voltage rejection figure of 96 dB

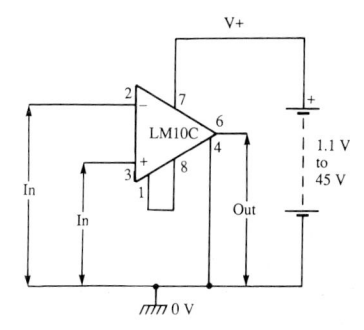

Figure 10.6 *Standard method of powering the LM10 from a single-ended supply*

172 LM10 op-amp/reference circuits

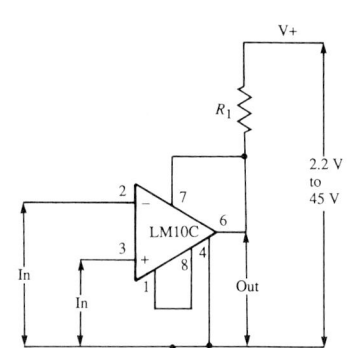

Figure 10.7 *Shunt method of powering the LM10 gives some unique circuit characteristics*

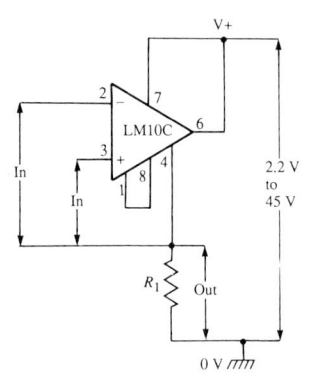

Figure 10.8 *An alternative shunt method of powering the LM10*

and wide operating voltage range, enable the LM10 to operate in either the linear or the switching mode while at the same time using its own output to modulate its own supply voltage and current!

Thus, this shunt mode of operation can be used in 2-wire remote-sensor applications in which the two wires carry both the supply current and the resulting signal information. Note that the minimum supply voltage used must be significantly greater than the normal 1.1 V value, to enable reasonable data amplitudes to be developed across R_1 without reducing the LM10 voltage below its minimum working value.

The reference amplifier

The built-in precision 200 mV reference and reference amplifier of the LM10 (see *Figure 10.2*) give the device great versatility and enable it to be used in many precision comparator and voltage regulator applications, etc. If this reference facility is not needed, or is to be used simply as a 200 mV reference, strap pins 1 and 8 of the IC together as shown in *Figure 10.9*. This gives the reference amplifier something useful to do, and makes a 200 mV 0-to-3 mA reference available between pins 1 and 4.

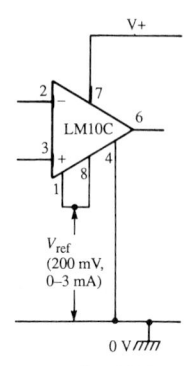

Figure 10.9 *Connection for obtaining a fixed 200 mV output reference from the LM10*

To get a precision reference in the range 0 to 200 mV, strap pins 1 and 8 together and then wire a 1k0 pot between pins 1 and 4 and take the output voltage from between pin-4 and the pot slider, as shown in *Figure 10.10*.

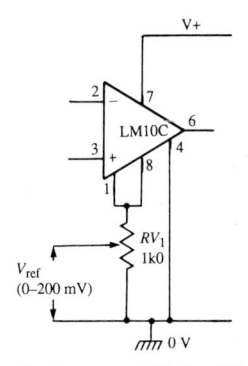

Figure 10.10 *Connection for obtaining a variable 0 to 200 mV output reference from the LM10*

174 LM10 op-amp/reference circuits

To get a precision reference in the range 200 mV and 40 V, use the connections shown in *Figure 10.11*. In this configuration the reference amplifier is used as a non-inverting dc amplifier with a fixed input of 200 mV and a voltage gain of $(R_1 + R_2)/R_2$.

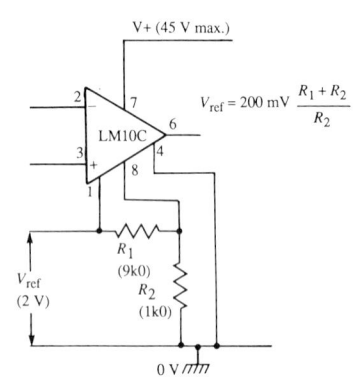

$$V_{ref} = 200\text{ mV}\ \frac{R_1 + R_2}{R_2}$$

Figure 10.11 *This connection enables any output reference in the range 200 mV to about 40 V to be obtained from the LM10*

Note that the reference amplifier has a typical unity gain bandwidth of about 500 kHz, and can thus be used as an ac amplifier in some special applications. Alternatively, it can be used as a simple voltage comparator by using the connections of *Figure 10.12*.

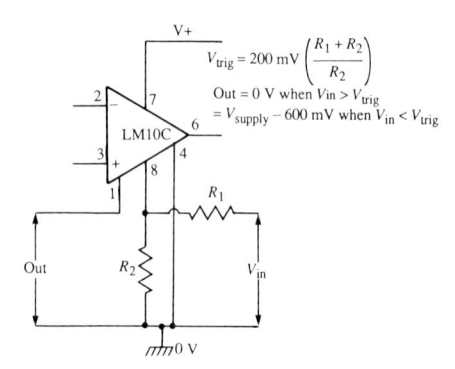

$$V_{trig} = 200\text{ mV}\left(\frac{R_1 + R_2}{R_2}\right)$$

Out = 0 V when $V_{in} > V_{trig}$
    = V_{supply} − 600 mV when $V_{in} < V_{trig}$

Figure 10.12 *This circuit enables the built-in reference amplifier to be used as a simple voltage comparator*

Op-amp applications

The op-amp section of the LM10 can be used in a wide variety of basic configurations, and *Figures 10.13* to *10.18* show a few of those that can be used when the IC is powered in the single-supply mode. *Figure 10.13* shows the basic connections for using the op-amp as an inverting dc amplifier. Note here that, since the op-amp output has a quiescent value of zero when its input voltage is zero, this circuit can only usefully accept input signals that are negative with respect to the zero volts rail.

Figure 10.14 shows how the above circuit can be modified to accept positive dc input signals. R_3 and R_4 are used to apply a positive bias voltage to the non-

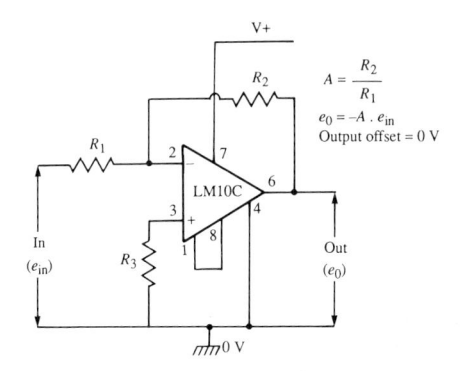

$$A = \frac{R_2}{R_1}$$
$$e_0 = -A \cdot e_{in}$$
Output offset = 0 V

Figure 10.13 *Basic inverting dc amplifier*

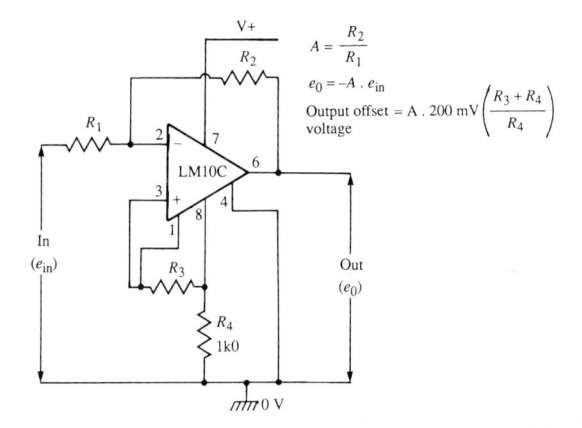

$$A = \frac{R_2}{R_1}$$
$$e_0 = -A \cdot e_{in}$$
Output offset = $A \cdot 200\,\text{mV} \left(\frac{R_3 + R_4}{R_4} \right)$ voltage

Figure 10.14 *Method of offsetting the output of the inverting dc amplifier*

176 LM10 op-amp/reference circuits

inverting terminal of the op-amp, so that its output takes up a positive 'offset bias' value when the circuit's input voltage is zero.

Figure 10.15 shows the connections for making an inverting ac amplifier. The output is biased to a quiescent value of half-supply volts (for maximum undistorted signal swing) via the R_3–R_4 divider, and the ac voltage gain is determined by the R_2/R_1 ratio. The input signal is ac coupled to R_1 via C_1.

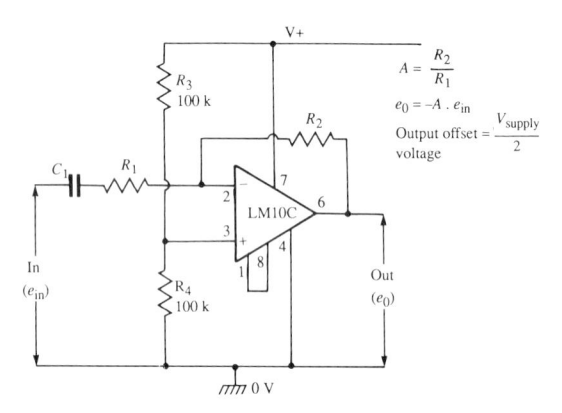

$$A = \frac{R_2}{R_1}$$
$$e_0 = -A \cdot e_{in}$$
$$\text{Output offset voltage} = \frac{V_{supply}}{2}$$

Figure 10.15 *Basic inverting ac amplifier*

Figure 10.16 shows how to use the LM10 as a non-inverting dc amplifier that will accept input signals down to ground volts. It can be used as a precision unity-gain voltage follower by simply removing R_1 and replacing R_2 with a short circuit.

Finally, *Figures 10.17* and *10.18* show standard methods of applying offset adjustment or compensation to the op-amp, using the LM10's built-in reference amplifier.

LM10 transmitter circuits

A major feature of the LM10 is its ability to be powered in the shunt mode shown in *Figure 10.7* in which, as already described, both the op-amp supply current and signal current flow through R_1, thus enabling the circuit to be used in 2-wire remote-sensor applications in which the op-amp is used as a remote transmitter that is coupled to the power-supply-plus-R_1 receiver unit via only two wires, as shown (for example) in the basic remote-amplifier circuit of *Figure 10.19*.

Here, the circuitry to the left of the dotted line acts as the remote transmitter, and is configured as a non-inverting ac amplifier with its voltage

LM10 op-amp/reference circuits   177

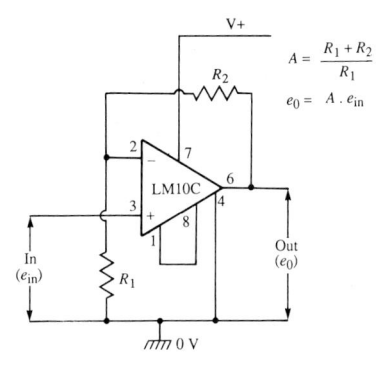

$$A = \frac{R_1 + R_2}{R_1}$$
$$e_0 = A \cdot e_{in}$$

Figure 10.16   *Basic non-inverting dc amplifier*

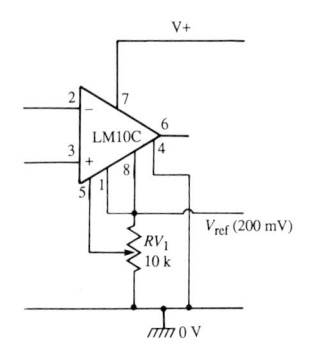

V_{ref} (200 mV)

Figure 10.17   *Standard method of offset adjustment or compensation*

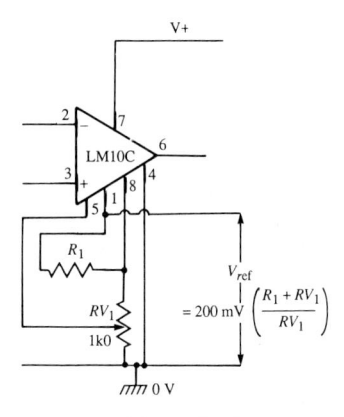

$V_{ref} = 200 \text{ mV} \left(\frac{R_1 + RV_1}{RV_1} \right)$

Figure 10.18   *Offset adjustment with boosted reference*

178 LM10 op-amp/reference circuits

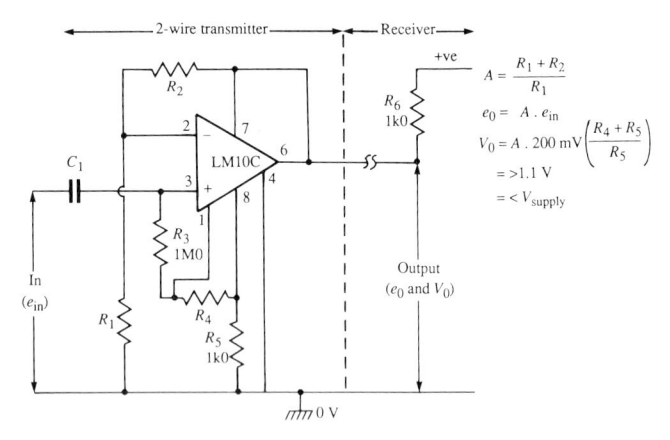

Figure 10.19 *Shunt-connected non-inverting ac amplifier or 2-wire transmitter*

gain set via R_1 and R_2 and with its output dc biased (via R_4 and R_5) to a quiescent value half-way between the positive supply value and the 1.1 V minimum operating voltage of the IC, thus enabling the largest possible undistorted output signals to be generated.

This transmitter is connected to the receiver (to the right of the dotted line) via only two wires (or via only a single wire if a common earth return is used), and its output signal is generated across R_6. This 2-wire type of circuit thus offers a very simple and inexpensive way of remote-monitoring a microphone or vibration sensor, etc. Some practical examples of such circuits are shown later in this chapter.

Comparator circuits

The fact that the LM10 incorporates both a precision voltage reference and a high-performance op-amp enables the device to be used in a variety of conventional and shunt-connected voltage comparator applications. *Figures 10.20 to 10.24* show examples of some basic circuits of these types.

Figure 10.20 shows the basic circuit of a fixed-value voltage comparator in which the output switches high when the input exceeds 200 mV. Here, pins 1 and 8 of the IC are shorted together to generate a 200 mV reference that is fed to the pin-2 non-inverting input terminal, and the circuit action is such that the op-amp output is low (at near-zero volts) when the pin-3 input voltage is less than the 200 mV reference value, and switches high (to full positive supply rail value) when the input exceeds the 200 mV reference value; the circuit thus acts as an over-voltage switch.

LM10 op-amp/reference circuits   179

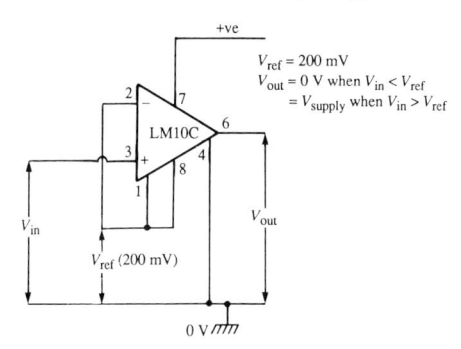

Figure 10.20 *Basic fixed-value (200 mV) voltage comparator*

$V_{ref} = 200$ mV
$V_{out} = 0$ V when $V_{in} < V_{ref}$
$= V_{supply}$ when $V_{in} > V_{ref}$

If desired, the circuit's action can be reversed (so that it acts as an undervoltage switch that gives a high output voltage when the input is below 200 mV) by simple transposing the pin-2 and pin-3 connections. Alternatively, it can be made to act as a general-purpose voltage comparator with a reference in the range 200 mV to 40 V by using the connections of *Figure 10.21*, in which the reference voltage value is set by the relative values of R_1 and R_2.

Figure 10.22 shows how the above circuit can be wired in the shunt or 2-wire transmitter mode, so that an input voltage can be easily monitored from a remote point.

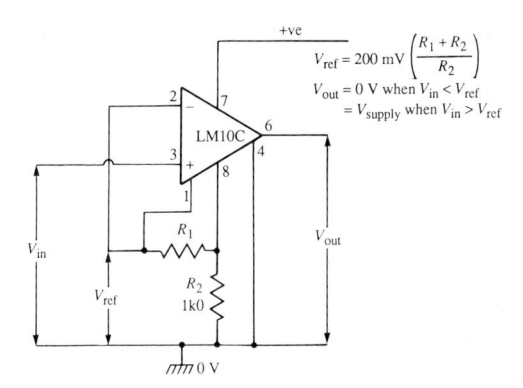

$V_{ref} = 200$ mV $\left(\dfrac{R_1 + R_2}{R_2} \right)$
$V_{out} = 0$ V when $V_{in} < V_{ref}$
$= V_{supply}$ when $V_{in} > V_{ref}$

Figure 10.21 *General-purpose voltage comparator*

Figure 10.23 shows how the LM10 can be used as a precision resistance comparator, in which the output switches high when the R_4 value exceeds that of R_3 (which can range from about 100 Ω to 10 Ω). Here, the R_1 to R_4 network is wired as a Wheatstone bridge and is powered via the precision 200 mV

180 LM10 op-amp/reference circuits

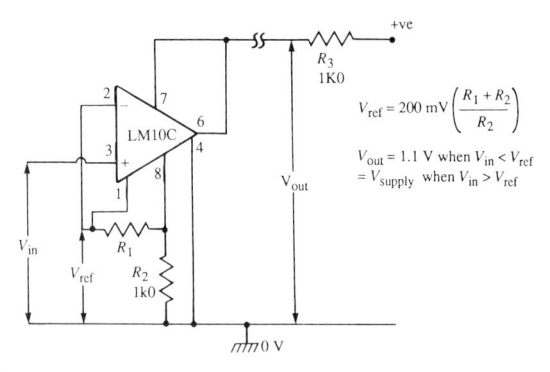

$V_{ref} = 200 \text{ mV} \left(\dfrac{R_1 + R_2}{R_2} \right)$

$V_{out} = 1.1 \text{ V when } V_{in} < V_{ref}$
$= V_{supply} \text{ when } V_{in} > V_{ref}$

Figure 10.22   *Shunt-connected voltage comparator*

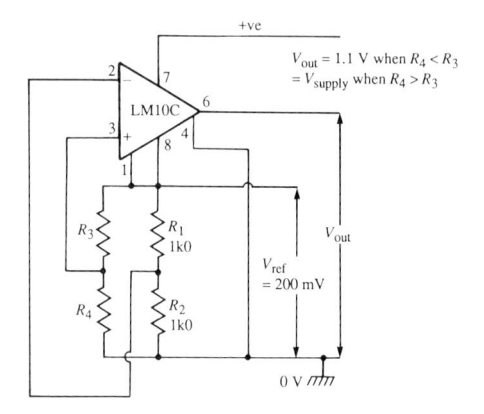

$V_{out} = 1.1 \text{ V when } R_4 < R_3$
$= V_{supply} \text{ when } R_4 > R_3$

Figure 10.23   *Basic resistance comparator circuit*

output of pin-1. The R_1–R_2 junction is fed to the op-amp's inverting input terminal, and the R_3–R_4 junction is fed to its non-inverting input terminal to give the necessary comparator action. Note that the circuit sensitivity can be increased by raising the reference voltage above its basic 200 mV value, but that the output current of the reference must not exceed 3 mA.

Figure 10.24 shows how this comparator can be used in the shunt or 2-wire transmitter mode (note in this case that the reference value should not exceed 1 V). If desired, R_4 (or R_3) may take the form of a thermistor or a light-dependent resistor (LDR) or some other type of resistive transducer, thus enabling this circuit to remote-monitor temperature or light levels, etc.

LM10 op-amp/reference circuits 181

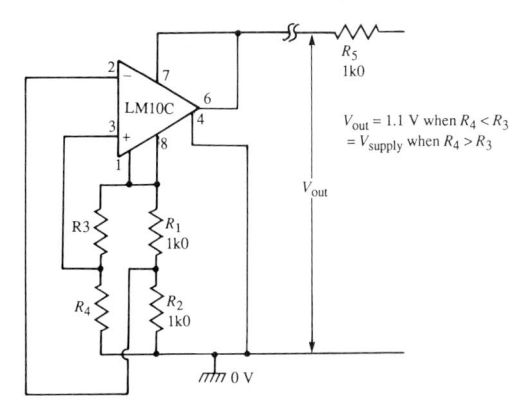

Figure 10.24 *Shunt-connected resistance comparator*

$V_{out} = 1.1$ V when $R_4 < R_3$
$= V_{supply}$ when $R_4 > R_3$

Astable circuits

The LM10 can be used in a variety of waveform generating applications, and *Figures 10.25* to *10.27* show some simple ways of using it in the astable multivibrator or square wave generator mode.

Figure 10.25 shows the basic astable circuit. This is a simple development of the standard dual supply op-amp astable, with R_1 and R_2 acting as a potential divider that sets the common point of the R_3-R_4 divider and the C_1-R_5 timing networks at half-supply volts. Because of the poor slew-rate of the LM10 the

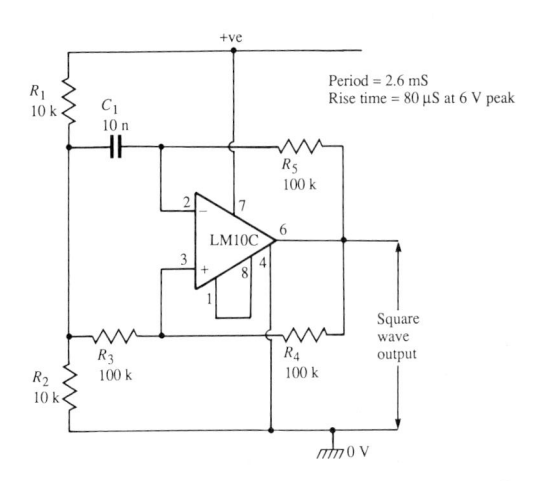

Period = 2.6 mS
Rise time = 80 μS at 6 V peak

Figure 10.25 *Basic astable multivibrator, with typical component values*

182 LM10 op-amp/reference circuits

circuit gives a fairly poor square wave output, with typical rise and fall times of about 80 μs when used with a 6 V supply. The circuit is, nevertheless, very useful in low frequency applications (up to a couple of kHz) as a simple alarm-tone generator or LED flasher, etc.

Figures 10.26 and *10.27* show alternative ways of gating the above astable on and off via an external control signal. Note that gate resistor R_6 must have a value that is small relative to the R_6 timing resistor. Several practical applications of these circuits are shown later in this chapter.

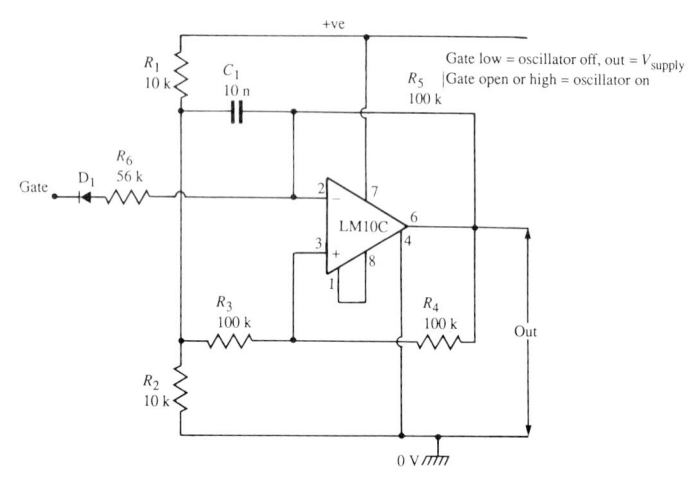

Figure 10.26 *Gated astable multivibrator*

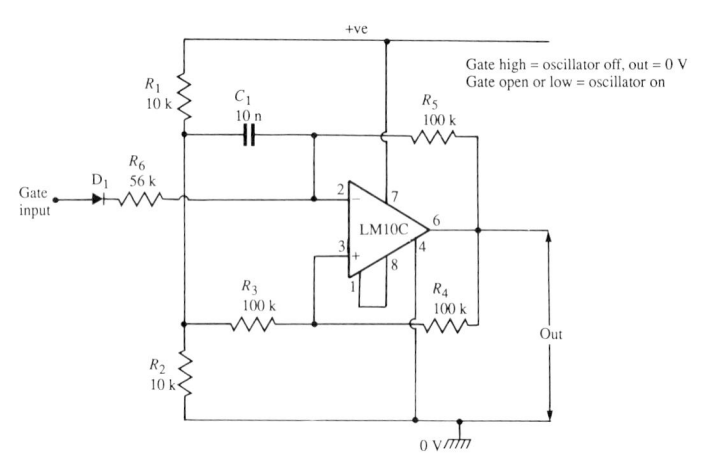

Figure 10.27 *An alternative version of the gated astable*

Voltage regulator circuits

The LM10 is, because of its built-in precision voltage reference and op-amp, suitable for use in a variety of voltage regulator applications, and *Figures 10.28* to *10.35* show a few practical circuits of this type.

Figure 10.28 shows the circuit of a low-power (up to 20 mA output) precision 200 mV to 20 V voltage regulator. Here, R_1 and RV_1 help generate a precision 200 mV to 20 V reference that is fed directly to the non-inverting input terminal of the op-amp, which acts as a unity-gain voltage follower and boosts the available output current to about 20 mA.

Figure 10.29 shows how to modify the circuit so that its output can be varied

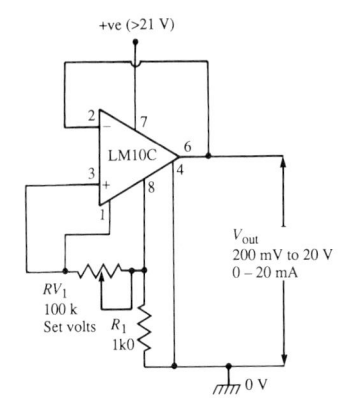

Figure 10.28 *Precision 200 mV–20 V voltage regulator*

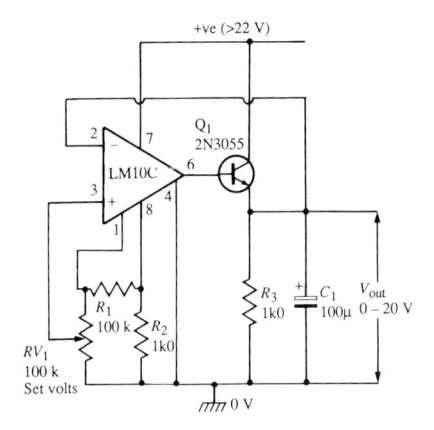

Figure 10.29 *Precision 0–20 V regulator with boosted output*

184 LM10 op-amp/reference circuits

all the way down to zero volts and its output current can be boosted to several hundred milliamps. Here, R_1–R_2 generate a fixed 20 V on pin-1 of the IC and across RV_1. The op-amp and power transistor Q_1 are configured as a composite unity-gain voltage follower, which directly boosts the 0–20 V output of RV_1 to current levels of up to several hundred milliamps.

An alternative approach to regulator design is shown in *Figures 10.30* and *10.31*, in which the op-amp is configured as a non-inverting × 25 amplifier. In the *Figure 10.30* circuit the op-amp input is taken directly from the fixed 200 mV reference, to give an unbuffered 5 V output. In the *Figure 10.31* circuit the

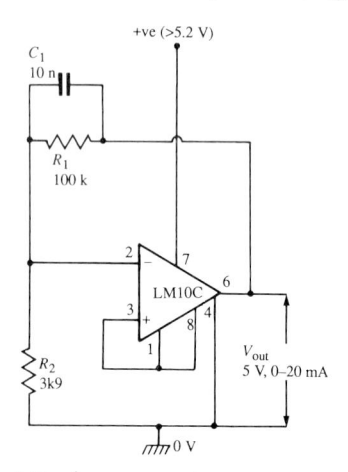

Figure 10.30 *Precision 5 V reference generator*

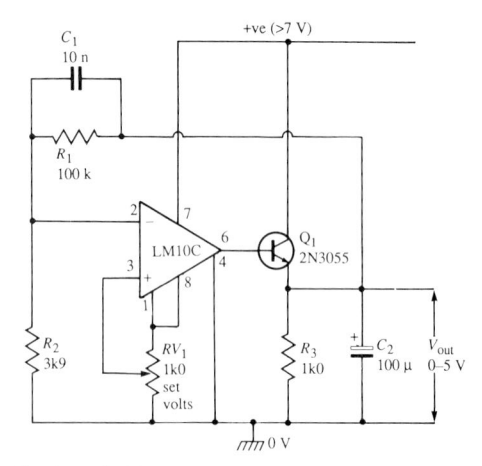

Figure 10.31 *Precision 0–5 V regulator*

LM10 op-amp/reference circuits 185

input is fully variable from zero to 200 mV via RV_1 and the available output current is boosted via Q_1, thus giving a high-current 0 to 5 V output. *Figures 10.32* and *10.33* show how the LM10 can be used in the floating

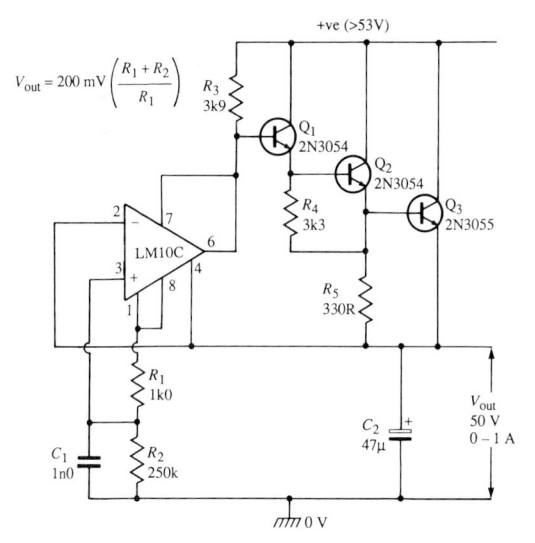

Figure 10.32 *Precision 50 V regulator*

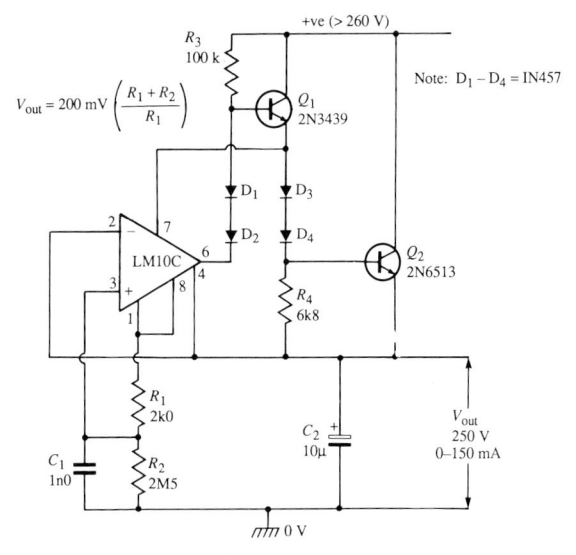

$V_{out} = 200 \text{ mV} \left(\dfrac{R_1 + R_2}{R_1} \right)$

Figure 10.33 *Precision 250 V regulator*

mode (in which pin-7 operates at less than the full supply voltage value and pin-4 operates at above the zero-volts value), to generate high output voltages. Note in both of these circuits that the IC is used in the shunt mode, with load resistor R_3, and that only a few volts are developed across the LM10 itself. In *Figure 10.32* the volt drop is limited to less than 2 V by the series-connected base-emitter junctions of Q_1 to Q_3. In *Figure 10.33* the drop is limited to a similar value via D_3–D_4 and the Q_2 base-emitter junction.

Figure 10.34 shows a simple example of the use of the LM10 as a low current (up to 20 mA) shunt-type 5 V regulator, and *Figure 10.35* shows how the IC

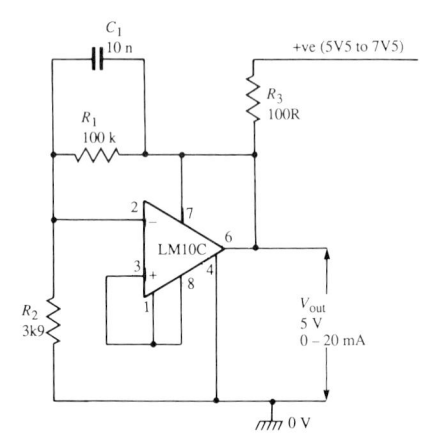

Figure 10.34 *A 5 V shunt regulator*

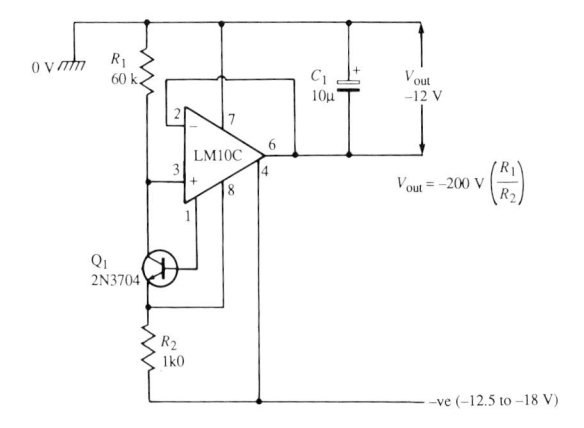

Figure 10.35 *A 12 V negative volt regulator*

can be made to act as a negative voltage regulator. In the latter case Q_1 is configured as a constant-current generator in which a precision 200 mV is developed across R_2, thus causing R_1 to pass a fixed current of 200 μA and thus generate a reference voltage of -12 V on input pin-3 of the op-amp, which is wired as a unity-gain voltage follower.

On/off-type voltage indicators

One major applications area of the LM10 is as an electronic fault indicator with an audible or visual output, and *Figures 10.36* to *10.43* show a variety of on/off-type fault indicator circuits in which the output is normally low (OFF) but goes high (ON) in the presence of a FAULT condition. In each of these circuits the IC's op-amp section is used as a simple voltage comparator with its output feeding to either a LED indicator or a low-power audible warning device via a suitable current-limiting (up to 20 mA) resistor.

In the *Figure 10.36* over-voltage indicator circuit the test voltage is fed to the non-inverting input terminal of the op-amp, and an internally-generated 200 mV to 40 V reference (set by R_1 and R_2) is fed to the inverting input pin, to set the circuit's trigger voltage value (V_{trig}). When the input test voltage is below V_{trig}, the op-amp output is low and the LED/alarm is off. When the input exceeds V_{trig}, the op-amp output switches high and activates the LED/alarm.

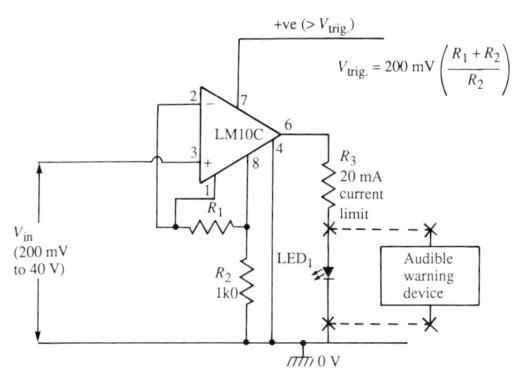

Figure 10.36 *Precision over-voltage indicator*

Figure 10.37 shows an under-voltage version of the above circuit, in when the output goes high when the input falls below a pre-set trigger voltage value. Here, the op-amp input terminal connections are simply transposed, with the input fed to the inverting pin and the reference fed to the non-inverting terminal.

188 LM10 op-amp/reference circuits

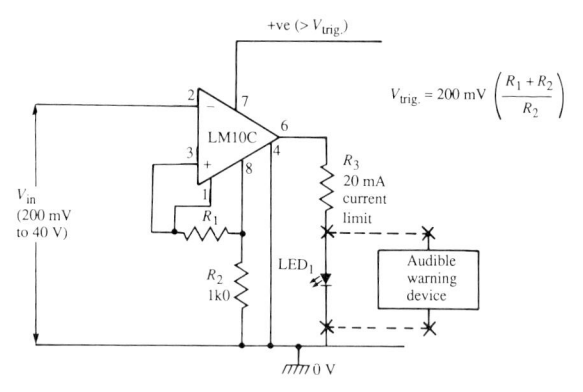

Figure 10.37 *Precision under-voltage indicator*

$V_{trig.} = 200 \text{ mV} \left(\dfrac{R_1 + R_2}{R_2} \right)$

Note that both of the above circuits give a very high input impedance, but need supply voltage values greater than the desired trigger voltage values.

Figures 10.38 and *10.39* show alternative voltage-indicating designs that do not suffer from these supply voltage restriction, and can operate from any supplies in the 2.5 to 40 V range. Here, the internal buffer amplifier is set to give unity gain and thus generates a fixed 200 mV reference voltage that is fed to one input terminal of the op-amp, and the test voltage is fed to the other input terminal via the R_1–R_2 potential divider, which thus determines the circuit's trigger voltage value.

In the *Figure 10.38* design the input test voltage is fed to the op-amp's non-inverting input terminal, so this circuit acts as an over-voltage indicator. In the *Figure 10.39* design the input is fed to the inverting terminal, so this circuit acts as an under-voltage indicator. Note that both of these circuits have a

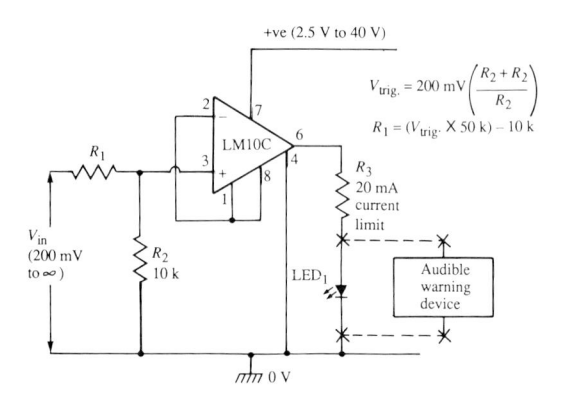

$V_{trig.} = 200 \text{ mV} \left(\dfrac{R_2 + R_2}{R_2} \right)$

$R_1 = (V_{trig.} \times 50 \text{ k}) - 10 \text{ k}$

Figure 10.38 *Alternative over-voltage indicator*

LM10 op-amp/reference circuits   189

basic input resistance sensitivity (equal to the sum of the R_1 and R_2 values) of 50 k/V. Thus, if the circuit is required to trigger at 12 V, R_1 needs a value of 12 × 50 k, minus 10 k, equals 590 k.

On/off-type current indicators

The *Figure 10.38* or *10.39* circuit can be given a maximum trigger voltage sensitivity of 200 mV by simple reducing the R_1 values to zero. Note in this case that the circuit will trigger when a current of 20 μA is passed through R_2, since 200 mV is developed across R_2 under this condition. These modified circuits can thus be used as current-sensitive indicators or switches. *Figures 10.40* and *10.41* show practical examples of such circuits. In each case, the R_2

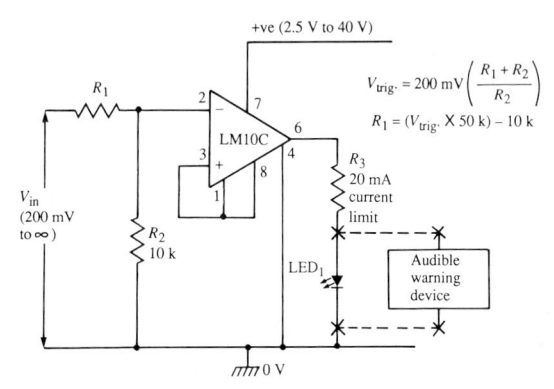

Figure 10.39   *Alternative under-voltage indicator*

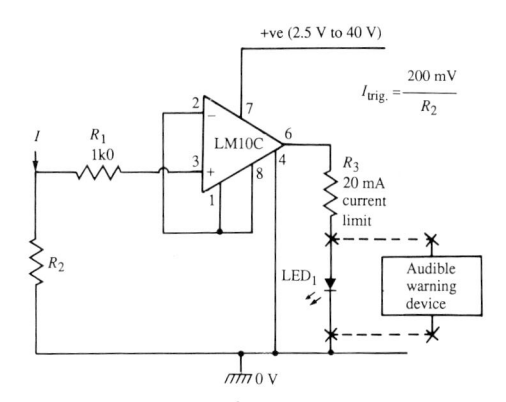

$$V_{trig.} = 200\,mV\left(\frac{R_1 + R_2}{R_2}\right)$$
$$R_1 = (V_{trig.} \times 50\,k) - 10\,k$$

Figure 10.40   *Precision over-current indicator*

$$I_{trig.} = \frac{200\,mV}{R_2}$$

190 LM10 op-amp/reference circuits

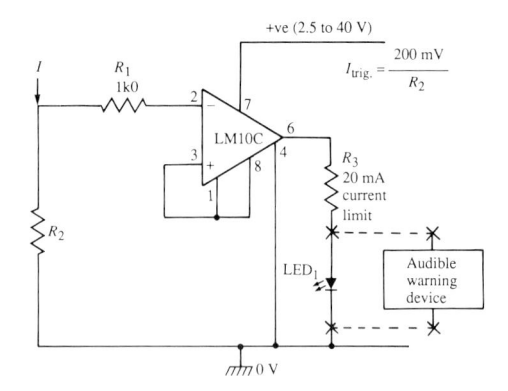

Figure 10.41 *Precision under-current indicator*

value is selected to give the required trigger current sensitivity (i.e., 1 Ω for 200 mA sensitivity, 10 Ω for 20 mA, etc.). Note that R_1 is used simply as a safety resistor, to limit op-amp input overload currents to a safe value.

On/off-type R indicators

Figures 10.42 and *10.43* show precision circuits that can be triggered by variations in the value of R_1, which would normally take the form of a

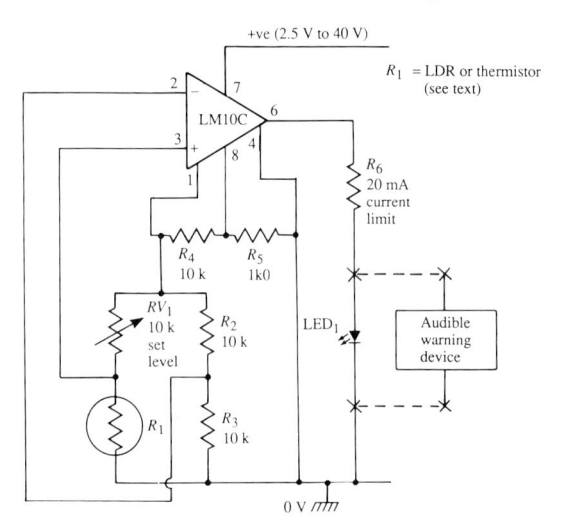

Figure 10.42 *Precision dark or under temperature indicator*

LM10 op-amp/reference circuits   191

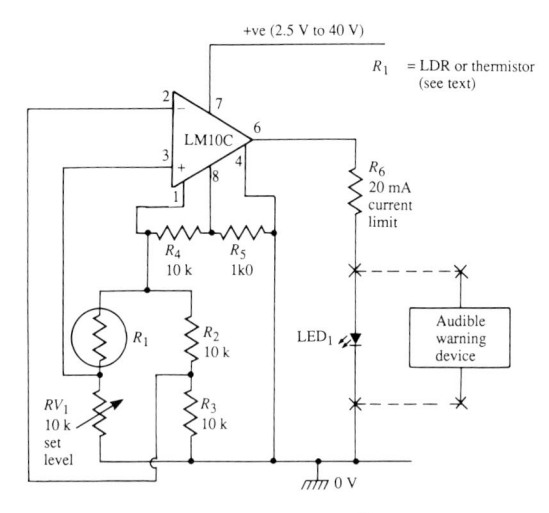

Figure 10.43 *Precision light or over temperature indicator*

resistive transducer such as a light-dependent resistor (LDR) or a thermistor, thus enabling the circuits to be triggered via either light or temperature. In such cases the LDR should take the form of a cadmium sulphide photocell, and the thermistor should be a simple negative temperature coefficient (NTC) type, each with a resistance value in the range 500 R to 9k0 at the required trigger level.

In each of these circuits the R_1 resistive element forms part of a Wheatstone bridge (formed by R_1–RV_1–R_2–R_3) that is powered (via pin-1) from the LM10's voltage reference amplifier, and the output of the bridge is used to activate the comparator-configured op-amp. The pin-1 voltage reference value is set at 2.2 V by the relative values of R_4–R_5.

Gated alarms

A particularly useful application of the LM10 is as a low voltage (3 V to 4.5 V) gated astable multivibrator, and *Figures 10.44* and *10.45* show how such a circuit can be made to act as either a gated LED flasher or as an audible alarm-tone generator.

The basic action of the gated astable was briefly described earlier. In essence, it acts like a standard dual supply op-amp astable, with R_1 and R_2 acting as a potential divider that sets the common point of the R_4–R_5 divider and the C_1–R_6 timing networks at half-supply volts. The circuit normally acts as a free-running square wave generator, but can be disabled (gated off) by

192 LM10 op-amp/reference circuits

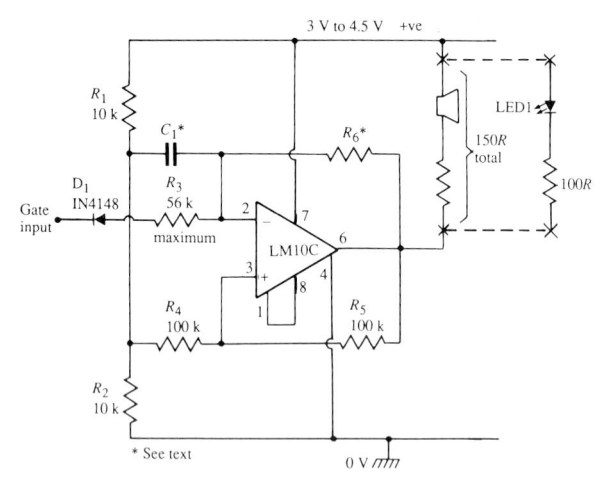

Figure 10.44 *LED flasher or alarm-tone generator that is gated on by a high (logic-1) input*

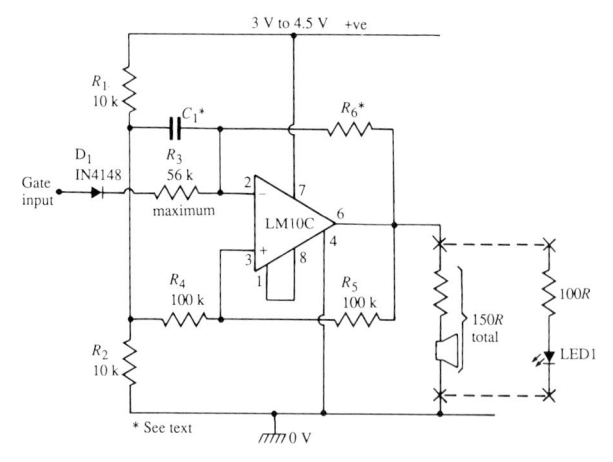

Figure 10.45 *LED flasher or alarm-tone generator that is gated on by a low (logic-0) input*

preventing its pin-2 terminal from exceeding the pin-3 value via the gate input terminal.

Thus, the *Figure 10.44* circuit is normally off (with its output locked high) but can be gated on by a high or logic-1 input, and the Figure 10.45 circuit is normally off (with its output locked low) but can be gated on by a low or logic-0 input.

The timing of these circuits is controlled by the C_1 and R_6 values. If a circuit is to be used as a LED flasher, these components should be given values of 220 n and 1M0 respectively, to give a flash rate of 100 flashes/minute. If a circuit is to be used as a 400 Hz alarm tone generator, the components should be given values of 10 n and 68 k respectively.

Gated fault indicators

Figures 10.46 to *10.49* show how the basic *Figure 10.44* and *10.45* gated astable circuits can be modified to act as fault condition indicators that give either a flashing LED or a 400 Hz monotone alarm signal output under the fault condition.

In the *Figure 10.46* over-voltage and *Figure 10.47* under-voltage alarm circuits the LM10's internal reference amplifier is used as a precision voltage comparator, with its trigger voltage (equals 4 V with the component values shown) determined by the relative values of R_5 and R_7, and its output is used to gate the astable via diode D_1.

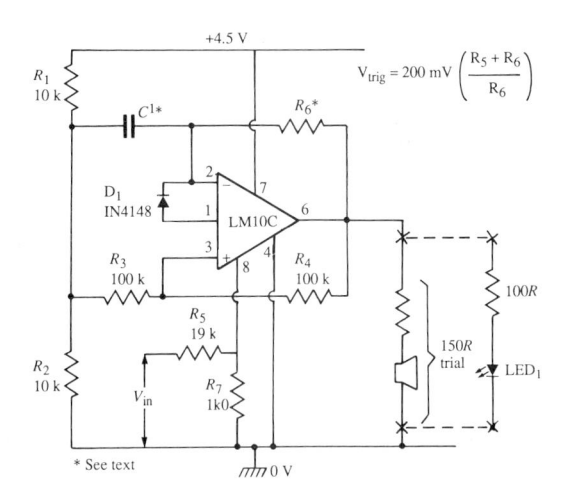

Figure 10.46 *Precision over-voltage alarm (triggers at 4 V with the R_5 value shown)*

In the *Figure 10.48* dark or under-temperature and *Figure 10.49* light or over-temperature alarm circuits the astable is gated via D_1 and the R_1–RV_1 potential divider, in which R_1 takes the form of either a cadmium sulphide photocell or an NTC thermistor with a resistance value in the range 500 R to 9k0 at the required trigger level.

194 LM10 op-amp/reference circuits

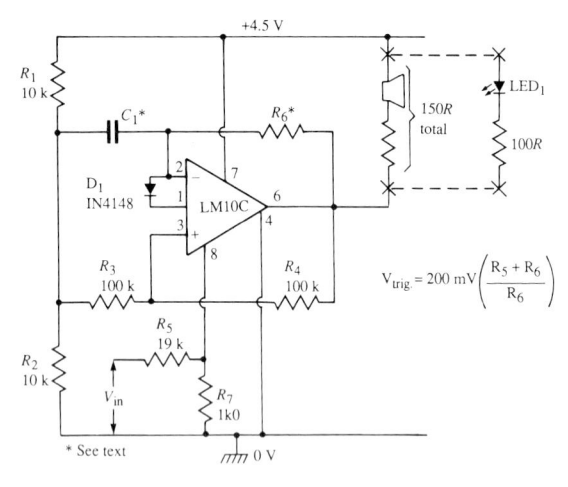

Figure 10.47 *Precision under-voltage alarm (triggers at 4 V with the R_5 value shown)*

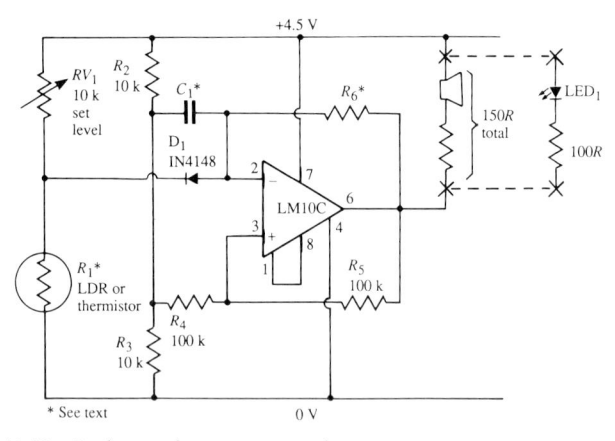

$$V_{trig.} = 200 \text{ mV}\left(\frac{R_5 + R_6}{R_6}\right)$$

Figure 10.48 *Dark or under-temperature alarm*

Note in these circuits that the astable timing is controlled by C_1 and R_6. These need values of 10 nF and 68 k to give a 400 Hz audible alarm tone output, or 220 nF and 1M0 if a circuit is to be used to give a LED-driving output at a flash rate of 100 flashes per minute.

Remote-amplifier circuits

One of the most interesting uses of the LM10 is as a shunt-connected 2-wire remote-amplifier transmitter of the type already shown in basic form in *Figure 10.19*. *Figures 10.50* to *10.52* show some practical examples of circuits of this type.

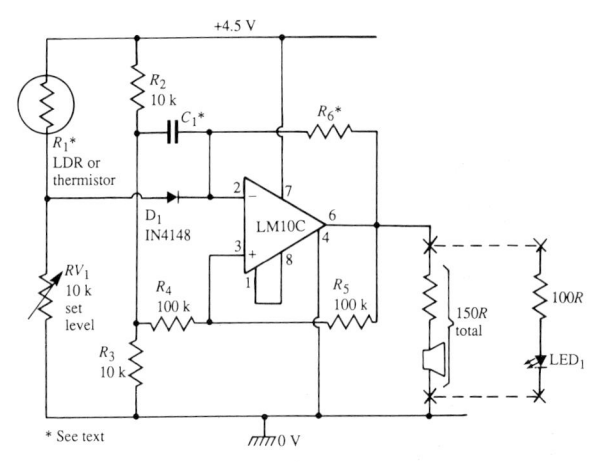

Figure 10.49 *Light or over-temperature alarm*

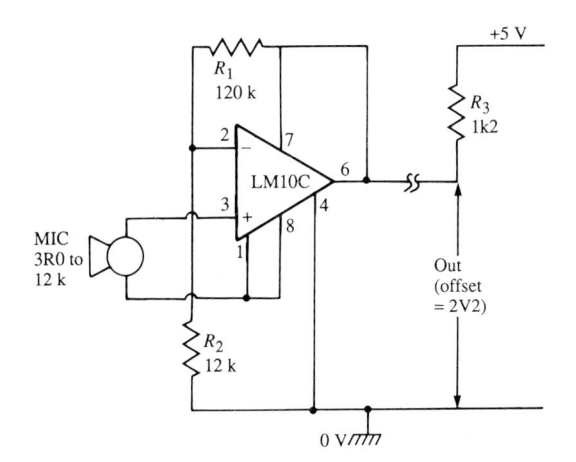

Figure 10.50 *Remote 20 dB voltage amplifier for use with inductive or magnetic input device*

196 LM10 op-amp/reference circuits

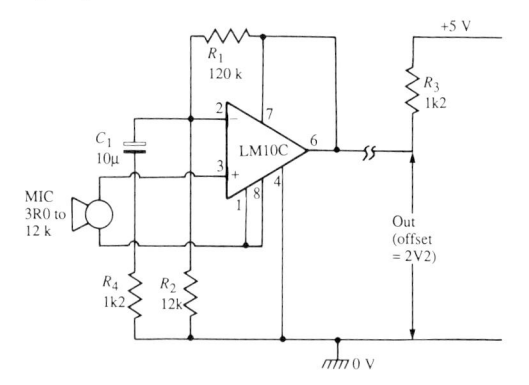

Figure 10.51 *Remote 40 dB voltage amplifier for use with inductive or magnetic input device*

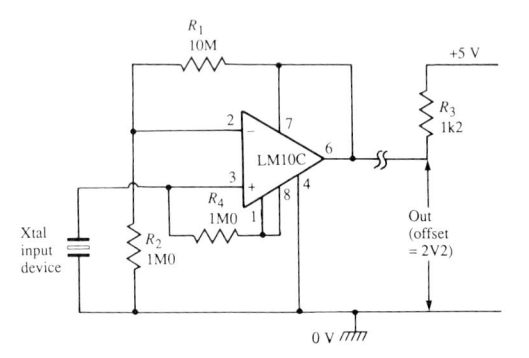

Figure 10.52 *Remote 20 dB voltage amplifier for use with a high-impedance (crystal) input device*

In these circuits the op-amp's positive supply and output terminals are shorted together, and both its supply and signal currents flow via R_3. The circuit to the left of R_3 thus acts as a signal amplifying transmitter that is connected to the R_3 'receiver' via only two wires. This 2-wire type of circuit thus offers a simple and inexpensive way of remote-monitoring a microphone or vibration sensor, etc.

Note in these circuits that pins 1 and 8 are shorted together so that a 200 mV reference voltage is generated and fed to the pin-3 non-inverting input terminal of the IC. Each circuit is given a dc voltage gain of × 11 via the R_1 and R_2 values and thus generates a quiescent output of 2.2 V, which can be modulated by the op-amp's output signal voltages.

The *Figure 10.50* and *10.51* circuits are suitable for use with low to medium

LM10 op-amp/reference circuits 197

input-impedance transducers such as moving coil or magnetic microphones, etc., and the *Figure 10.52* circuit is suitable for use with high impedance devices such as crystal or ceramic microphones, etc. The *Figure 10.50* and *10.52* designs each give an ac signal voltage gain of about 20 dB, and the *Figure 10.51* design gives a gain of 40 dB.

Figure 10.53 shows a simple modification that allows the basic design to be used with any resistive transducers in the R_x position. The op-amp signal gain is set at × 7.6 via R_1 and R_2, but the pin-3 input voltage is variable from 200 mV upwards via R_x, thus giving the output voltages shown in the diagram.

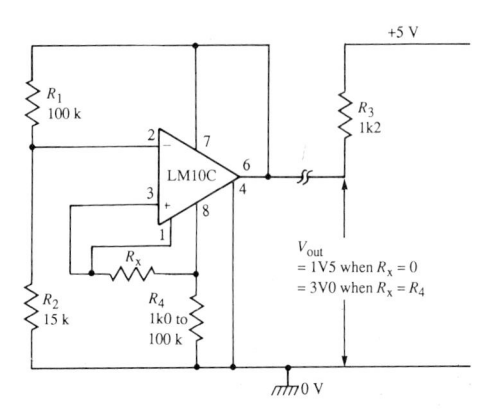

Figure 10.53 *A 2-wire transmitter for use with a variable-resistance sensor*

Fault indicators

The shunt-connected 2-wire amplifier technique can easily be modified to form the basis of a variety of 2-level fault indicator transmitters with either resistor, LED or transistor outputs at their receiver ends, as shown in *Figures 10.54 to 10.57*.

In each of these circuits the LM10 is wired as a precision voltage comparator with a fixed reference voltage applied to one input terminal and a variable input applied to the other, and with its output grounded via R_3 and its supply current passing through R_4. The basic circuit action is such that when the op-amp output (pin-6) is low (giving zero drive into R_3) only 200 mV is lost across R_4, but when the output is high (giving a high drive current into R_3) about 3.5 V are lost through R_4.

In practice, the circuit output can either be taken directly from the pin-7 supply terminal of the IC, or can be taken via a LED or a transistor that is wired across R_4 as shown dotted in the diagram. In the former case the LED

198 LM10 op-amp/reference circuits

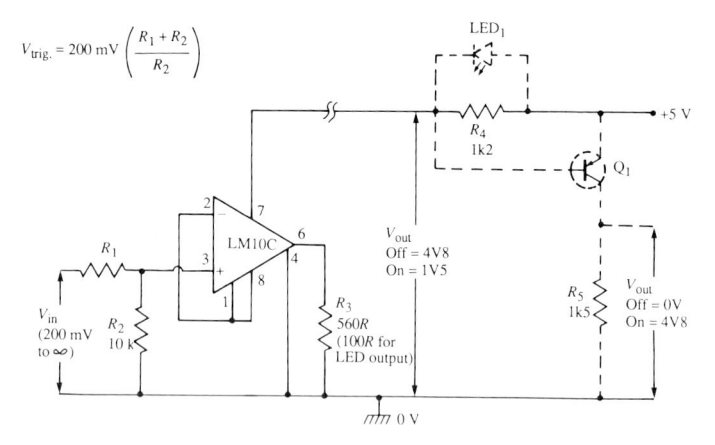

Figure 10.54 *2-wire precision over-voltage transmitter with resistor, LED or transistor output*

'ON' current is determined by the R_3 value. In the latter case the final output can be taken from across Q_1 collector load resistor R_5 and switches between an OFF value of zero volts and an ON value of 4.8 V.

In each of the *Figure 10.54* to *10.57* circuits pins 1 and 8 are shorted together, to generate a 200 mV reference voltage that is applied to one of the input pins of the op-amp. In *Figure 10.54* the input voltage is fed to the pin-3 non-inverting terminal of the op-amp via potential divider R_1–R_2, and the

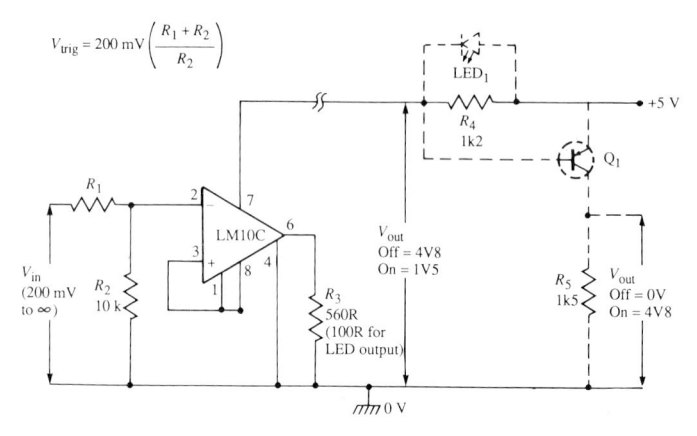

Figure 10.55 *2-wire precision under-voltage transmitter with resistor, LED or transistor output*

circuit action is such that pin-7 output switches low (to 1.5 V) when the pin-3 voltage exceeds 200 mV; this circuit thus acts as an over-voltage detector. The *Figure 10.55* circuit gives the reverse of this action, with its pin-7 output going low when the pin-3 input voltage falls below 200 mV, and thus acts as an under-voltage detector.

The *Figure 10.56* circuit is similar to that of *Figure 10.54* except that the pin-3 input voltage is taken from across R_1 and is thus directly proportional to the R_1 input current. This circuit thus acts as an over-current detector. The *Figure 10.57* circuit is similar, but has its input terminals reversed, and thus acts as an under-current detector.

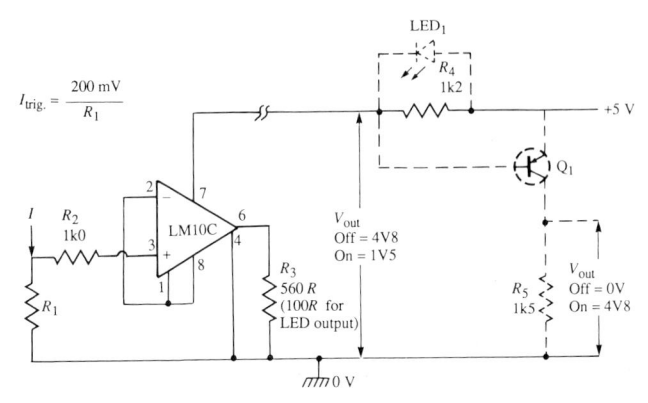

Figure 10.56 *2-wire precision over-current transmitter with resistor, LED or transistor output*

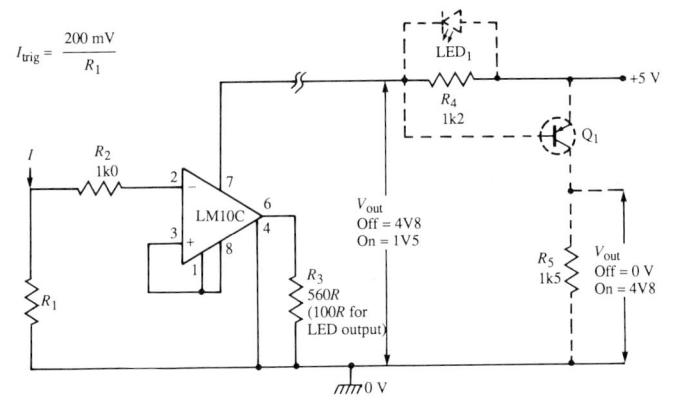

Figure 10.57 *2-wire precision under-current transmitter with resistor, LED or transistor output*

Figure 10.58 shows how this type of circuit can be modified to act as a light- or temperature-sensitive detector by using a cadmium-sulphide photocell or NTC thermistor in the R_1 position. In this case R_1–RV_1–R_2 and R_3 are wired in the form of a Wheatstone bridge that has its outputs fed to the pin-2 and pin-3 inputs of the op-amp, and the bridge is powered from a 400 mV supply set up via R_4 and R_5.

The LDR/thermistor should have a resistance in the range 500 Ω to 9k0 at the desired trigger level. If the LDR/thermistor is wired as shown the circuit will act as either a dark or under-temperature detector, its output switching on when either the light level or the temperature falls below a value pre-set via RV_1. Alternatively, the circuit can be made to act as a 'light' or 'over-temperature' detector (in which the output switches on when the light or temperature level rises above a pre-set value) by simply transposing the positions of R_1 and RV_1.

Fault alarms

The *Figure 10.54* to *10.58* fault-detecting circuits give a simple ON or OFF form of fault indicating output. By contrast, *Figures 10.59* to *10.62* show a selection of fault detectors that give either a flashing LED or a 400 Hz audible monotone alarm signal output and can thus be fairly described as fault alarm

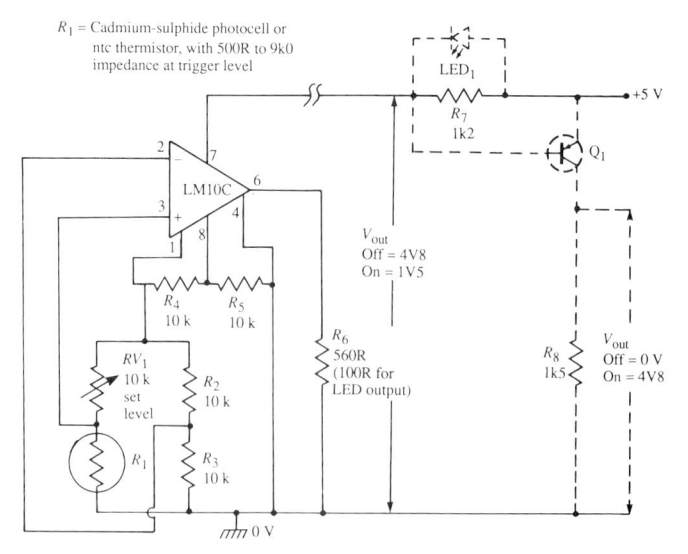

Figure 10.58 *2-wire precision dark or under-temperature transmitter with resistor, LED or transistor output*

LM10 op-amp/reference circuits 201

circuits. These alarms are in fact very similar to the four gated fault indicator circuits described in *Figures 10.46* to *10.49*, except that they are connected in the 2-wire shunt mode and can thus give remote indication of the fault condition.

In each of the *Figure 10.59* to *10.62* circuits the op-amp section of the LM10 is wired as an astable multivibrator or square wave generator that is gated (via D_1) via the output of the internal reference amplifier, which is wired as a voltage comparator with one input taken from the 200 mV band-gap reference and the other taken from an external source via pin-8. The *Figure 10.59* and *10.60* circuits are wired as under-voltage and over-voltage alarms respectively, and *Figures 10.61* and *10.62* are wired as under-current and over-current alarms respectively.

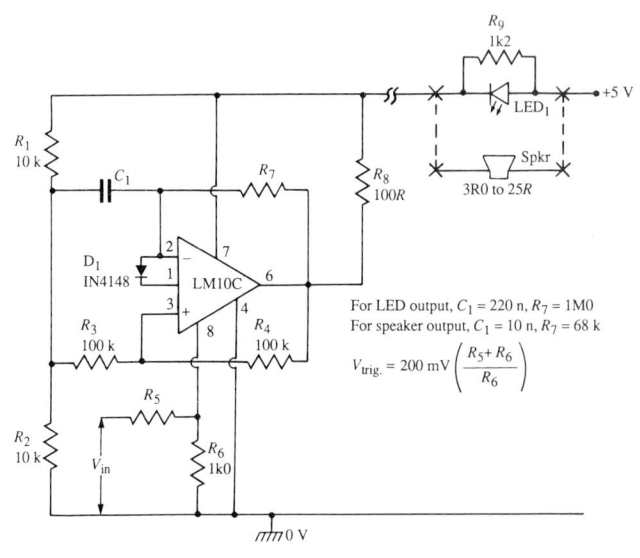

For LED output, $C_1 = 220$ n, $R_7 = 1M0$
For speaker output, $C_1 = 10$ n, $R_7 = 68$ k

$$V_{trig.} = 200 \text{ mV} \left(\frac{R_5 + R_6}{R_6} \right)$$

Figure 10.59 *2-wire precision under-voltage transmitter with flashing LED or monotone audio (400–Hz) output*

C_1 and R_7 control the astable timing of these circuits. They need values of 10 nF and 68 k to give a 400 Hz audible alarm output, or 220 nF and 1M0 to give a 100-flashes-per-minute LED-driving output.

Meter amplifier circuits

To conclude this, look at the LM10, *Figures 10.63* to *10.66* show a variety of ways of using the device as a very-low-voltage amplifier that can be used to

202 LM10 op-amp/reference circuits

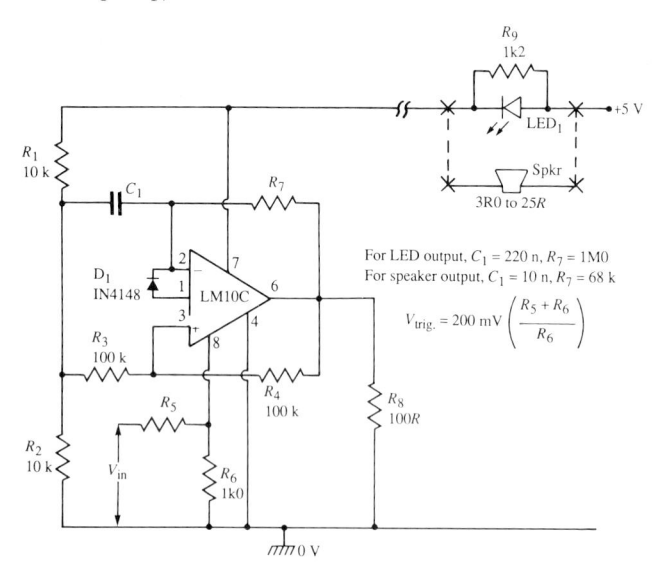

Figure 10.60 *2-wire precision over-voltage transmitter with flashing LED or monotone audio (400-Hz) output*

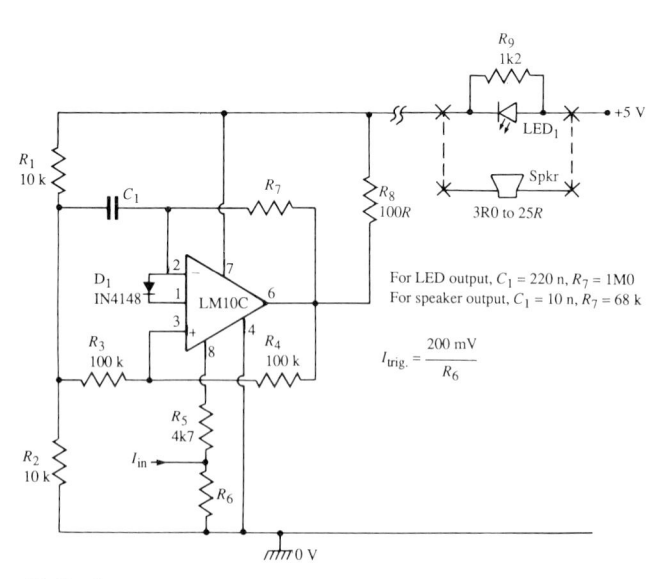

For LED output, $C_1 = 220$ n, $R_7 = 1\text{M}0$
For speaker output, $C_1 = 10$ n, $R_7 = 68$ k

$$V_{\text{trig.}} = 200 \text{ mV} \left(\frac{R_5 + R_6}{R_6} \right)$$

Figure 10.61 *2-wire precision under-current transmitter with flashing LED or monotone audio (400-Hz) output*

For LED output, $C_1 = 220$ n, $R_7 = 1\text{M}0$
For speaker output, $C_1 = 10$ n, $R_7 = 68$ k

$$I_{\text{trig.}} = \frac{200 \text{ mV}}{R_6}$$

greatly increase the effective sensitivity of a normal moving-coil meter or multimeter.

In the *Figure 10.63* circuit the op-amp is used as a simple non-inverting amplifier and increases the effective sensitivity of a 1 mA meter by a factor of about 100, to 10 μA fsd. Note that this circuit has no 'set null' facility, and can give no indication of reverse-connected signals. The modified circuit of *Figure 10.64* does not suffer from this defect.

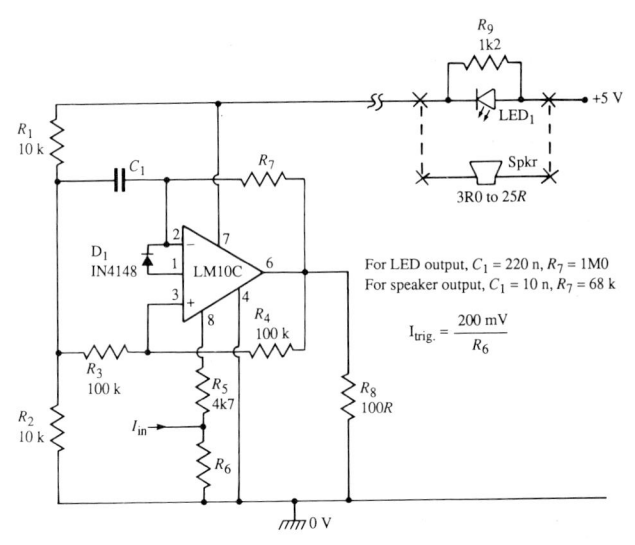

For LED output, $C_1 = 220$ n, $R_7 = 1M0$
For speaker output, $C_1 = 10$ n, $R_7 = 68$ k

$$I_{trig.} = \frac{200 \text{ mV}}{R_6}$$

Figure 10.62 *2-wire precision over-current transmitter with flashing LED or monotone audio (400-Hz) output*

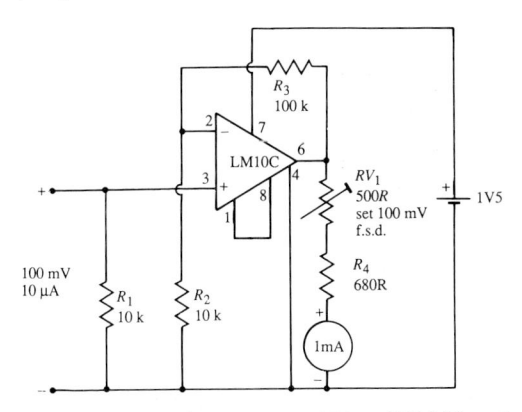

Figure 10.63 *Simple meter amplifier giving a sensitivity of 100 k/V to a 1 mA fsd meter*

204 LM10 op-amp/reference circuits

The *Figure 10.64* circuit uses a 100 µA meter and increases its effective sensitivity by a factor of 100, to 1 µA fsd. Note in this case that the 200 mV reference voltage of pin-1 is used to create a common input and output signal line that is 200 mV above the zero volts value, thus enabling the op-amp output to swing between +1.3 V and −0.2 V and enabling the meter to indicate both forward and reverse voltage values. RV_1 provides the circuit with a null facility, enabling the output to be set to give a meter reading of zero with zero input signal applied.

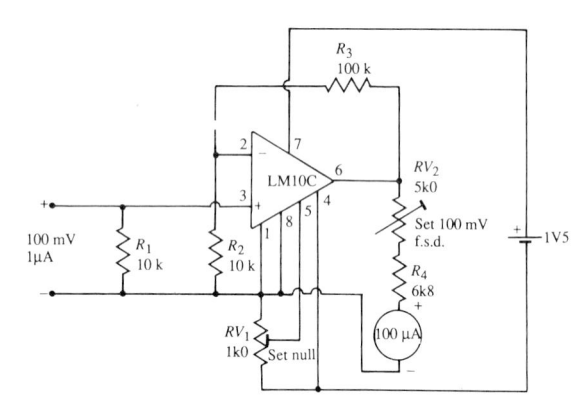

Figure 10.64 *Precision meter amplifier, with null facility, giving a sensitivity or 1M0/V to a 100 µA meter*

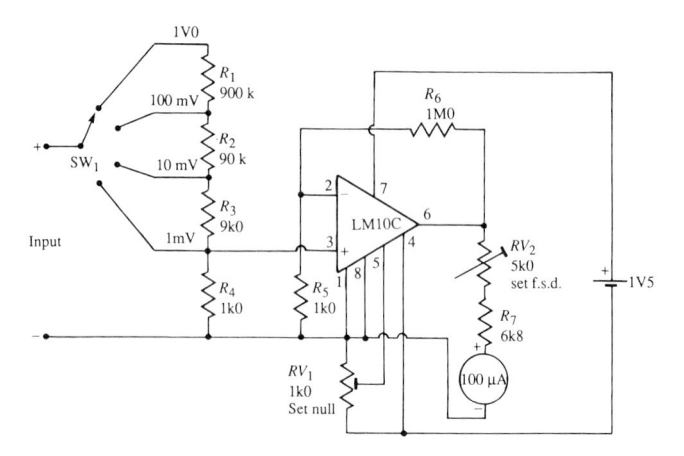

Figure 10.65 *A 4-range dc millivoltmeter, with a sensitivity of 1M0/V*

LM10 op-amp/reference circuits   205

Finally, *Figures 10.65* and *10.66* show how the basic *Figure 10.64* circuit can be adapted for use as a 4-range dc millivolt meter and a 4-range dc microammeter respectively. Note that each of these op-amp driven meter amplifier circuits is powered from a single 1.5 V battery supply.

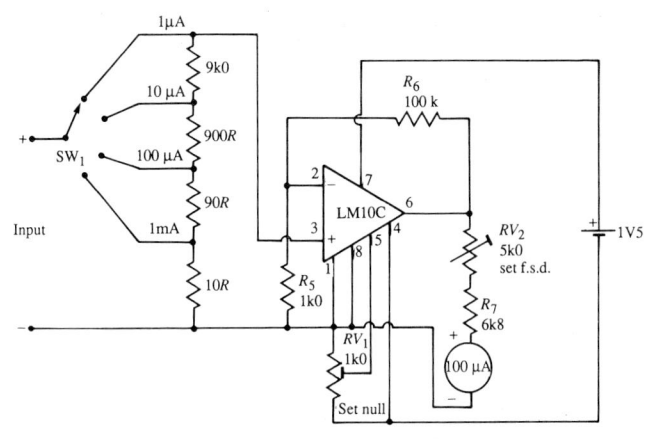

Figure 10.66   *A 4-range dc microammeter, with a full scale sensitivity of 10 mV*

Index

AC amplifier, inverting/non-inverting 13 Fig.
AC/DC converters 71-2, 73 Fig.
Active filters 30-1, 32 Fig.
Active filter circuits 31-4
Active tone control circuit 36 Fig.
Adders 28-9
Adjustable voltage power supply 16 Fig.
Adjustable voltage reference 15 Fig.
Analogue-activated circuits 44-5, 46-7 Figs.
Analogue meter circuits 75-8
Analogue subtractor (differential amplifiers) 14 Fig.
Applications, round up 12
Audio mixer (inverting analogue adder) 14 Fig.

Balanced phase splitter 29-30
Basic configurations 3-5
Basic symbol 3 Fig.
Biasing accuracy 25-6
Bidirectional DC voltage follower with boosted output current drive 27 Fig.
Bridge-balance detector/switch 16 Fig.

CA 3080 OTA circuits 130-45
  closed-loop operation 137-8
  comparator circuits 141-2
  offset balancing 138-9
  Schmitt trigger circuits 142-3
  voltage-controlled gain 139-41

208 Index

Closed-loop amplifiers 5–6
Compound circuits 84–101
Compound inverting amplifiers 91–3
Compound relaxation oscillators 93–5
Compound voltage followers 86–90
Current-booster follower circuits 26–7
Current boosting 84–6

DC millivoltmeter circuit 75 Fig.
Differential amplifier (analogue subtractor) 14 Fig.
Differential voltage comparator 5 Fig.
Diode stabilization 55–6
Dual op-amp types 10 Fig.
DVM converter circuits 72–5

Electronic rectifiers 68–9

Five-range linear-scale ohmmeter 78
Frequency response of curve of 741 op-amp 4 Fig.

Half-wave rectifier circuit 69 Fig.
High-gain open-loop AC amplifier 4 Fig.
High-pass second-order active filter 14 Fig.
High-value regenerative over-voltage switch 41 Fig.
High-value under-voltage switch 41 Fig.

Input biasing of op-amp 25, 26 Fig.
Inverting amplifier circuits 19–20
Inverting analogue adder (audio mixer) 14 Fig.

LM10 op-amp and reference circuits 168–205
    astable circuits 181–2
    comparator circuits 178–9
    fault alarms 200–1
    fault indicators 197–200
    gated alarms 191–3
    gated fault indicators 193–4
    meter amplifier circuits 201–5
    on/off-type current indicators 189–90

Index 209

on/off-type R indicators 190–1
on/off-type voltage indicators 187–9
op-amp applications 175–6
power supply connections 170–1
reference amplifier 173–4
remote-amplifier circuits 195–7
shunt operation 171–2
transmitter circuits 176–7
voltage regulator circuits 183–7
LM13600 OTA circuits 146–67
  automatic gain control amplifier 156–7
  controlled impedance buffers 149–50
  linearizing diodes 146–9
  offset biasing 156
  VCA circuits 152–6
  voltage-controlled filters 160–4
  voltage-controlled oscillators 164
  voltage-controlled resistors 157–60
LM13700 IC 167

Manually-triggered bistable 66 Fig.
Micro-power operation 47–9

Non-inverting amplifier circuits 20–3
Norton op-amp circuits 102–29
  biasing techniques 106
  comparators 110–4
  current mirrors 104–5
  current regulator circuits 116–8
  I_{set} programming 125–7
  Schmitt circuits 110–2
  voltage regulator circuits 115–6
  waveform generator circuits 119–21
  widebank amplifiers 128–9

Offset nulling 12, 13 Fig.
Op-amp basics 2
Op-amp parameters 7–8

Peak detector with buffered output 69 Fig.
Practical op-amps 9–12
Practical transmitter circuits 98–100
Precision half-wave AC/DC converter 17 Fig.
Precision half-wave rectifier 17 Fig.
Precision light-activated oscillator/alarm 63 Fig.
Precision over-temperature oscillator/alarm 63 Fig.
Precision rectifiers 70–1

Quad op-amp types, parameter/outline details 11 Fig.

Ramp-rectangle generator with variable slope-mark/space ratio 66 Fig.
Relaxation oscillator circuit 59 Fig.
Resistance activation 61–4

Sample-pulse generator 49–51
Schmitt trigger 67 Fig.
Sine/square function generator 18 Fig.
Sine wave oscillators 52–4
Single-supply manually-triggered bistable 67 Fig.
Speech filter 34, 36 Fig.
Square wave generators 59–60, 61 Fig.
Subtractors 28–9
Supply connections of op-amp 3 Fig.
Supply-line splitter 15 Fig.
Switching circuits 65–7

Thermistor stabilization 54–5
Three-band active tone control circuit 37 Fig.
Triangle/square generation 64–5, 66 Fig.
Twin-T oscillators 57–9
Two-wire information systems 95–7
Two-wire receiver 100–1

Variable active filters 34–5
Variable over-voltage switch with regenerative feedback 14 Fig.
Variable symmetry 61, 62 Figs.
Voltage comparator circuits 38–51

Voltage follower circuits 23–4, 25 Figs.
Voltage reference circuits 78–9
Voltage regulator circuits 80–3

Waveform generator circuits 52–67
Wien network 52–4
Window comparators 42–3